Remembering Lattimer

THE WORKING CLASS IN AMERICAN HISTORY

A list of books in the series appears at the end of this book.

Remembering Lattimer

Labor, Migration, and Race in Pennsylvania Anthracite Country

PAUL A. SHACKEL

UNIVERSITY OF ILLINOIS PRESS
Urbana, Chicago, and Springfield

1 2 3 4 5 C P 5 4 3 2 1
∞ This book is printed on acid-free paper.

Library of Congress Cataloging-in-Publication Data
Names: Shackel, Paul A., author.
Title: Remembering Lattimer : labor, migration, and race in Pennsylvania anthracite country / Paul A. Shackel.
Description: Urbana : University of Illinois Press, [2018] | Series: The working class in American history | Includes bibliographical references and index.
Identifiers: LCCN 2017061622| ISBN 9780252041990 (hardcover : alk. paper) | ISBN 9780252083686 (pbk. : alk. paper)
Subjects: LCSH: Anthracite Coal Strike, Pa., 1897. | Labor movement—Pennsylvania—History. | Working class—Pennsylvania—History.
Classification: LCC HD5325.M62 S53 2018 | DDC 331.89/282233509748 32—dc23
LC record available at https://lccn.loc.gov/2017061622

E-book ISBN 978-0-252-05073-2

CONTENTS

PREFACE

The Anthracite Heritage Project has been working in northeastern Pennsylvania since 2009, initially with an exploration of the meaning and memory of the Lattimer massacre. In later years, the project has focused on labor and the everyday lives of the immigrants and their families, who came to the region to work in the coal mines. For this manuscript I am indebted to Dr. Kristin Sullivan who, as a graduate student, made some initial contacts with the Hazleton community. Kristin built a Web page that outlined some of the history associated with the Lattimer massacre and demonstrated the University of Maryland was interested in doing further research into the event's history. Though the event does not reside in the national public memory, it was clear that many people in northeastern Pennsylvania were determined to preserve and protect a memory of Lattimer. Within days Kristin found several enthusiastic residents, including Joe Michel and John Probert, who were more than willing to share their love for the region's history. Both Joe and John, longtime residents of the area, provided a wealth of material and an entrée into the community. They told us about their relationship to the event and they spoke about some of the long-held stories told by locals. John provided our initial introduction to local history and introduced us to many community members. Joe gave us access to "the 110," his collection of documents and maps related to the region. As coal-mining companies closed their operations, they often abandoned their buildings and left behind company records. Many records have been lost to the wrecking ball. However, whenever possible, Joe salvaged documents, and they have become the foundation for our research into the everyday life and heritage of northeastern Pennsylvania mining communities. It was Joe's enthusiasm for the project that gave me, Dr. Michael Roller, and Justin Uehlein access to his archives for our initial research into Lattimer. Because of Joe's generosity, his material became the foundation for Michael's dissertation and the background context for this book.

Dr. Michael Roller led the archaeology project at the Lattimer massacre site. Together, we spoke with Pasco Schiavo, a prominent lawyer in the region and owner of the property where the massacre occurred. He generously permitted us to perform an archaeological survey on the property. Mike also contacted Dan Sivilich of BRAVO (Battlefield Restoration and Archaeological Volunteer Organization), and Dan generously responded with his time, as well as that of his crew. Through freezing temperatures, a stiff breeze, and a light snow, Dan's crew braved two weekends that made the archaeological survey an overwhelming success. Though not part of this Lattimer memory project, Camille Westmont has participated in the larger Anthracite Heritage Project, collecting data on the general conditions of patch town life in northeastern Pennsylvania. Kyla Cools and Katie Boyle have also recently joined the project as we moved it to Eckley Miners' Village. I am proud of my students and former students, who I consider friends and colleagues. They have worked tirelessly to connect with the community and make the project much more accessible to the public. While they graduate and move on in their professional careers, their impact on this project and the local community will be long lasting. They were professionals at all times, and will continue to make a difference in the communities in which they work.

Throughout the entire project Angela Fierro, a life-long resident of Lattimer and owner of the Lattimer company store, was more than generous with her time and resources. She allowed me and my students access to family histories and other primary documents. During the survey of the Lattimer massacre site, she generously provided lunch and allowed us to warm up in her house. Angela was always a great resource, connecting us to people in the community. Most of the community is excited about our interest in their local heritage, although, at times, people did question our goals and perspectives.

There are many reporters working for the local newspaper who brought attention to the project and allowed us to share our story with the community. In particular, the *Standard Speaker* covered almost every aspect of our project. So, thank you, Bobby Masso and Jim Dino, who wrote the first stories about our project. I am especially grateful to Kent Jackson, who has written many subsequent stories about the Anthracite Heritage Project, emphasizing our program's goal of examining the historical and contemporary issues of work and immigration.

Though northeastern Pennsylvania has struggled with the acceptance of the new immigrant, a century ago as well as today, I admire the work of residents who strive to make the life of the new immigrant more bearable and to create a more harmonious community. I am inspired by the dedication of Joe Maddon, manager of the Chicago Cubs, who uses his resources and star power to bring attention to the needs of a fractured community, as well as the hard-working members of the Hazleton Integration Project staff, including Bob Curry and Elaine Maddon Curry, who have helped to recruit high school students from underrepresented groups for our project.

I am grateful to the many reviewers of this manuscript, including Dean Saitta and two anonymous reviewers. I also appreciate that Katie Boyle and James Engelhardt provided helpful comments on the manuscript. The manuscript is a better product because of their helpful input.

Remembering Lattimer

INTRODUCTION

Toward the middle of 1897, the UMW (United Mine Workers), later known as the UMWA (United Mine Workers of America), began a strategic push to enroll members in the union in the anthracite region of northeastern Pennsylvania. During several weeks of protest and strikes in mid-August and early September 1897, union leaders began organizing many of the foreign-born, unnaturalized workers in and around Hazleton, Pennsylvania, one of the largest coal industry and commercial support centers in the region. Ironically, earlier that year the UMWA was instrumental in convincing state legislators to pass an anti-immigrant bill that would tax employers for each non-U.S. citizen worker on their payroll. In turn, the coal companies deducted this tax from the workers' salaries.

The Hazleton area was at a boiling point over the tax and other perceived injustices implemented by the coal operators. Strikers walked off the job and sometimes forcefully closed mines. The county sheriff and his posse tried to contain the strike. However, on September 10, 1897, about 250 men from Harwood began a five-mile march to close the colliery at Lattimer, a coal-patch town north of Hazleton. By the time the protesters reached Lattimer, the demonstration had grown to about four hundred striking miners of Polish, Slovak, and Lithuanian descent. Without warning, the deputies fired into the unarmed crowd, killing nineteen miners and wounding about thirty-eight others. The men who died were all foreign born and unnaturalized. The incident is the most serious act of labor violence in the Commonwealth of Pennsylvania and one of the most troubling, yet forgotten, moments in U.S. history. The sheriff and his deputies stood trial, and after five weeks, were acquitted. As a result of the Lattimer trial, the miners realized the power of capital over labor, and they increasingly saw the importance of organizing to fight for economic justice. Lattimer strengthened the role of the UMWA in the anthracite coal region, as tens of thousands of foreign-born miners flocked to the union, despite the organization's initial anti-immigrant position. The incident at Lattimer convinced the union that foreigners could be organized, and it strengthened the UMWA's position. By 1900, under the leadership of John Mitchell, the UMWA became the fastest-growing union in the country, leading to successful strikes in 1900 and 1902.

There are lessons we can learn from the Lattimer tragedy. The story of Lattimer is about how science and public sentiment allowed for the racialization of foreigners, which justified their exploitation, and the sanctioning of their murders through either direct confrontation or through labor exploitation. Within the context of the events at Lattimer, we also can view the re-creation of these xenophobic fears and prejudices in our contemporary society.

The massacre has been largely forgotten in the national public memory. The beginning of this amnesia is probably related to the racial attitudes toward those slain and the efforts by those who controlled the anthracite coalfields to command the narrative of the event. It was beneficial for the coal barons to forget the conflict at Lattimer and the demands of the coal workers. Even though the Lattimer massacre is not part of the national public memory, there are local efforts, with various levels of success, to keep the memory of the event alive, along with the struggle to achieve social and economic justice. The story of Lattimer is a lesson about the past, and it is also about the present and the future.

I interviewed several current and former residents of the Hazleton area. The majority expressed their disappointment that the massacre does not receive much attention at the national level. They were also disillusioned that the Lattimer massacre is not part of the Commonwealth of Pennsylvania school curriculum. If local residents learn about the massacre, it is through family histories, and perhaps experiencing a local commemorative event. The incident at Lattimer is surprisingly absent in the literature related to U.S. labor history, even though in 1901, it was considered by some as one of the five greatest strikes in America (*Grand Forks Daily Herald*, September 8, 1901). Its impact on the national stage faded rapidly in the early twentieth century. When searching through history books that focused on labor strife, the most related historicized events include the Great Railroad Strike (1877), the McCormick Harvesting Machine Company Strike, which led to the Haymarket Riot (1886), Homestead (1892), Pullman (1894), the Anthracite Coal Strike (1902), the Bread and Roses Strike (1912), Ludlow (1913), the Patterson Silk Strike (1913), the Battle of Blair Mountain (1921), and the Textile Workers' Strike (1934), to name a few. There are many more, depending on one's background and perspective. However, the Lattimer strike is missing from our contemporary narratives of labor and ethnic history protests, despite the fact that the incident had one of the highest death tolls when compared to these other major strikes. Even in Howard Zinn's *A People's History of the United States: 1492 to Present* (Zinn 2003, 284), best known for his radical and political histories of the United States, the event is obscured and he only mentions Lattimer in passing when describing the sinking of the *USS Maine* in February 1898. He refers to the monthly journal of the International Association of Machinists, which explained that the explosion of the *Maine* was horrendous. However, the journal's editor noted that at the same time, people's senses are often dulled to the killing of innocent protesters, like those at Lattimer.

In 1997, I noticed that the Pennsylvania Historical and Museum Commission (PHMC) and the UMWA, among other organizations, were planning to hold commemorative events marking the centennial of the labor massacre. During the first day, I witnessed the unveiling of a state historic roadside marker on a highway near Harwood, Pennsylvania. Harwood is the community where the miners began their march toward Lattimer. There were a few actors dressed in period costume, one as Mother Jones, and a few as miners. Mother Jones was not at Lattimer in 1897, and she did not visit the area until the 1900 Anthracite Coal Strike. Participants unveiled the commemorative marker, with several dignitaries making speeches as they commemorated the beginning of the march to Lattimer. We hopped on a bus and drove slowly past Saint Stanislaus cemetery in Hazleton, where fourteen of the victims are buried. A commemorative stone marker on the wall of the cemetery could be viewed from the street. Then, we journeyed to the Lattimer memorial, where a new PHMC state-sponsored roadside marker was unveiled. A contingent of union members and labor leaders walked from St. Mary's, the Roman Catholic Church in Lattimer, to the site of the massacre. Cecil Roberts, president of the UMWA, among others, provided short speeches, followed by the unveiling of a roadside marker. The same costumed actors dressed in period clothes were present.

The next day the Wyoming Historical and Geological Society (now the Luzerne County Historical Society), with funding from the Pennsylvania Humanities Council, supported a symposium at Eckley Miners' Village, with many of the speakers discussing the events surrounding the Lattimer massacre. Several professional and amateur historians provided varying perspectives about the strike. Michael Novak, author of a popular historical novel based on the event, was the main speaker. The audience sparred and debated over interpretations of some of the primary and secondary documentation as well as the meaning of Lattimer today.

I believe the Lattimer event is a nationally significant story that needs to be made part of the national public memory. This monograph provides details about the Lattimer massacre using primary and secondary resources, and I show the varying perspectives and competing memories of Lattimer. Commemorative events often provide renewed interest at the local level; however, the memory of Latimer still falls short of the national collective memory. This monograph explores the ways that the event is remembered, forgotten, and sometime repressed. Importantly, the descendants of the late-nineteenth-century immigrants who were racialized and demeaned by the English-speaking community are now in a position of power. Logically, they have a connection to the contemporary Latino immigrants arriving in the region who have come in large numbers over the past two decades. However, rather than embracing differences and finding commonalities, some in the now traditional local population are treating the new immigrants with the same prejudices their ancestors received several generations ago. The city of Hazleton, which is known as a city of immigrants, has gone as far as creating local anti-immigration legislation, which was eventually struck down by the courts. The community is often showcased as the forerunner of the anti-immigration movement of the twenty-first century. The region is now a stronghold of the Tea Party, which had a major impact on the 2016 national elections.

Included in this historical overview and ethnography of this community is a summary of an archaeology project to locate the site of the Lattimer massacre. Archaeology, in itself, is a way of commemorating and making a place visible to a larger audience. Significant resources are invested in a project, and people begin to see the place as important and worthy of investigation, remembering, and commemorating. There is growing momentum in using archaeology to examine and highlight the history of labor and immigration by using archaeological techniques. For instance, the archaeology related to the Ludlow massacre in Colorado serves as a good example of how archaeologists explored issues related to labor concerns and living conditions of immigrant working families. Through the archaeology of the tent colony, the Ludlow Collective, a multi-institutional group of scholars and students, explored questions about the formation of temporary communities, protest labor movements, and government and military intervention. More important, the archaeology at Ludlow raised the visibility of this bloody episode in labor relations and is helping to make this incident part of the broader public memory (Ludlow Collective 2001, 94–107; McGuire and Reckner 2002, 44–58; Saitta 2007; Walker 2000, 58–75; Wood 2002). Ludlow is now a National Historic Landmark.

Being entrenched in the local northeastern Pennsylvania community, it became clear that many in the community are proud of their ethnic and immigrant heritage. Town festivals and church bazaars highlight the heritage of the historic immigrants, the Italians and Slavs, who came to northeastern Pennsylvania more than a century ago. At the turn of the twenty-first century, northeastern Pennsylvania has become home to a new wave of immigration—Latinos, mostly from the Dominican Republic. There are also newcomers from Mexico and other Latin American counties. In Hazleton, the Latino population is approaching 50 percent and rising quickly. Perhaps in a few years it will be a majority minority city. With the influx of these new residents, the local white population began to realize that their dominant presence, which controls the heritage narrative, is diminishing. Their heritage is quickly being subsumed by new social and cultural traditions. Latino stores and restaurants are now part of the Main Street landscape. It is easy to get a sample of Latin roots music in neighborhoods during the evening as people sit on their front porches and have discussions with neighbors next door or across the street. Tensions have developed over the new immigrants, and the city of Hazleton led the way in 2006 by legislating an anti-immigration law. These events made it increasingly clear to me that our project needed to focus on the historic struggle of immigrants, as well as the importance of connecting our work to the story of the contemporary immigrants in northeastern Pennsylvania as well as the United States.

The story of Lattimer is about immigration and how the new, undocumented residents of the coalfields were racialized and treated as inferior, disposable people. In 1978, Michael Novak of the American Enterprise Institute wrote a historical novel focusing on the Lattimer massacre. Novak was self-described as a Catholic scholar, and his writings are very much from a conservative perspective. He is an advocate of unrestrained capitalism and has questioned Pope Francis's comments about wealth redistribution. *The Guns of Lattimer* (1978) stands today as one of the most complete historical resources related to the Lattimer massacre. The book provides an accurate and detailed accounting of events. When the *Wall Street Journal* reviewed the book that year, the critic asked, "It is tempting to ask Mr. Novak why we really need the book. The incident occurred more than 80 years ago. It sounds like a unique event that would best be forgotten. Besides, American society has changed; American bosses don't act that way toward blue-collar workers anymore" (Wysocki 1978, 24). That last sentence, "American bosses don't act that way toward blue-collar workers anymore" is a haunting phrase. It seems to be a common strategy in neoliberal America—to take tragic episodes in labor history, think of them as lessons of the past, and not consider how ethnic inequalities and labor injustices are playing out in the world today. The book reviewer here is way off the mark, and he clearly did not want to address the realities of the social consequences of unchecked capitalism.

The message of Lattimer remains very relevant today in northeastern Pennsylvania, as well as the rest of the country. There are many connections between what we are uncovering about the social, economic, and political circumstances that surround historic Lattimer and the prejudices rampant in this community today. If we can commemorate the struggle of the immigrants who came here more than a century ago, I believe we also can appreciate the heritage and social values of the contemporary immigrant. Therefore, our challenge as heritage scholars who support social justice issues is to connect the heritage of the Italians and Slavs to the contemporary Latino immigrants. There are universal values that we all want and desire—such as peace, good health, education, and the ability to sustain oneself. Connecting these universal values with the new immigrants can make the Lattimer project valuable and relatable to any ethnic group not originally associated with the historic event. This demonstrates the power of heritage, as we struggle to find ways to use our history to promote social justice issues in contemporary America.

CHAPTER 1

Anthracite Mining

Background on Anthracite Mining

The incident at Lattimer and the story of European immigration into northeastern Pennsylvania begins with the discovery and eventual marketing of anthracite coal. About 600 million years ago, in a swampy environment, trees and other organic material fell into water, and the sparse supply of oxygen thwarted bacterial decomposition. The sea advanced over the coastal swamps, which then contained peat deposits. Sea sediments of sand, silt, and clay covered the peat. When the sea retreated, the pressure of the sediments forced moisture and poisonous gases out of the peat. Lignite, or brown coal, formed. Eventually, geological pressures along with the development of faults and folds in the earth transformed the coal into anthracite, a metamorphic rock. The creation of anthracite is the final product of the geological process known as coalification. Anthracite by definition contains more than 87 percent carbon. Though difficult to ignite, it burns longer and cleaner than any other type of coal (Rogers 1858, 970–1000; Wallace 1987, 7).

The anthracite coal region in Pennsylvania is located in the northeastern section of the state and covers about 484 square miles (figure 1.1). The area contains most of the world's supply of anthracite. The coal is located in several narrow bands divided into three fields—southern, middle (sometimes subdivided into the eastern middle and the western middle), and northern, and they run in a northeasterly direction. The bands are part of the northern portion of the Appalachian region. Most of the coal is found in seams, or "veins," that can be a few inches to as much as forty to sixty feet thick. The veins in the northern and middle fields are nearly horizontal or tilt at a moderate angle. The southern field has veins that

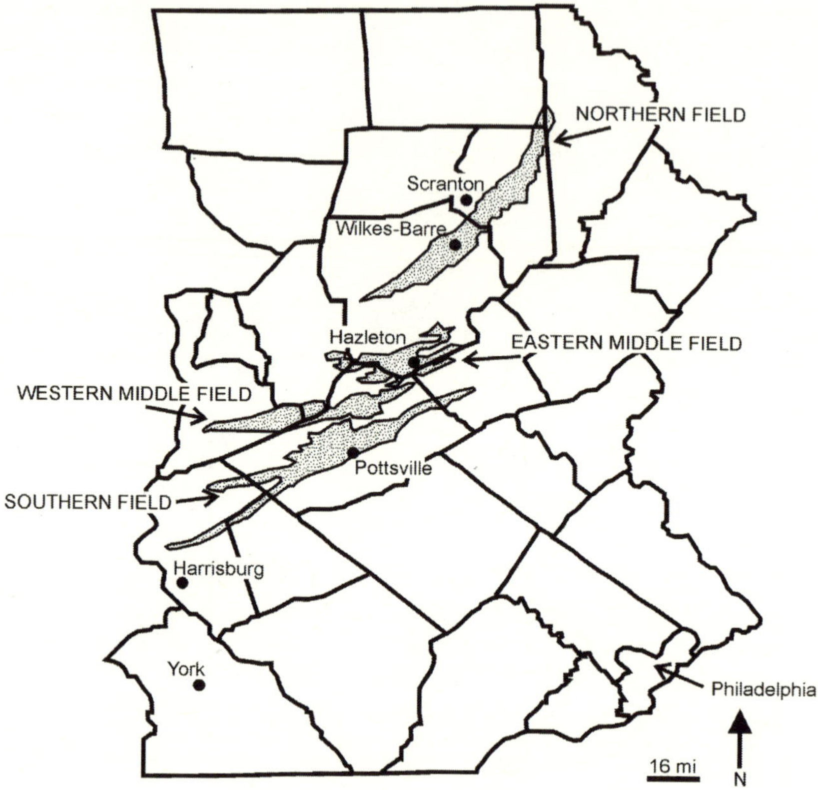

Figure 1.1. Map showing northeastern Pennsylvania, with major cities and the major anthracite coal fields: Northern Field, Western and Eastern Middle Fields, and the Southern Field. (Image drawn by V. Camille Westmont.)

have been subject to more dramatic lateral forces, creating undulating and dramatic folds. Erosion in the southern field is prominent, exposing some of the veins, making it initially easier to mine (Jones 1914, 4–5; Wallace 1987, 5).

Anthracite coal was first mined in the Wilkes-Barre region in northeastern Pennsylvania in 1775. Within a few decades, this resource was found farther south in the Schuylkill and Lehigh regions (Wallace 1987). Farmers or village artisans often operated the early mines. Unlike in other nineteenth-century industry, coal miners generally worked in small, unsupervised teams. The extraction effort tended to focus on the excavation of outcroppings on the land surface. This open-pit mining required low skill and little capital. The men would shovel the coal into wagons, where it would then be taken to local railheads, and then to canals. By the 1830s, many of these open-pit mines became exhausted, and more sophisticated mining techniques developed, like drift and slope mining. Drift mining is the process of creating a horizontal tunnel into a mountain coal seam and then following and mining the vertical coal seam. Slope mines require digging a tunnel down a slope and into a seam. These new techniques required greater engineering skills and capital investment than surface mining (Keil and Keil 2015, 6–7; Luther 1881, 48; Miller and Sharpless 1985, 85–86; Wallace 1987, 8). For more than a century, the process of extracting coal remained simple, without heavy machinery. In 1902, a journalist described, "The process of mining is simple, and tools are the rudest. They are pick and shovel, bar, hand and machine drills, the latter an auger, turned by a crank—and powder and squibs" (Rhone 1902, 56).

Anthracite extraction played a critical role in the development of the American industrial revolution. By the 1820s and 1830s, East Coast industries began to replace waterpower with coal. Beginning in the 1830s, coal also became an increasingly important source for residential heating and cooking. The annual output of Pennsylvania anthracite increased from 21,000 tons in 1830 to 3,300,000 tons in 1847. The growing iron industry east of the Alleghenies was fed by the increased output of coal, and as a result, iron became cheaper and more accessible to East Coast industries (Chandler 1972, 151, 156–58; MacGaffy 2013, 4).

Early American industrialists promoted an ideal of economic progress based on republican virtues. They believed that a harmony between employer and employees would protect republican society (Palladino 2006, 43–44). Skilled miners were scarce in the early development of the industry and the miners usually garnered the respect of the coal operators, who frequently worked alongside them in the mines. Companies attracted miners from England, Scotland, and Wales. These workers, skilled at their trade, accepted a level of industrial discipline, like quotas for output, in exchange for adequate housing and a free supply of coal. There was no great status difference between employer and employees. They shared similar ethnic and religious backgrounds and the belief in mutually beneficial economic growth. Workers and employers continued this relationship through the 1820s and 1830s. "The vision of 'republic' industry—or at least financial independence—seemed a real possibility" (Palladino 2006, 45–46; also see Wallace 1987, 177–79).

By the 1830s, three primary canal companies developed to transport anthracite coal to urban markets and seaports in the northeast. New York and

Philadelphia investors financed these operations: the Delaware & Hudson Canal Company, the Schuylkill Navigation Company, and the Lehigh Coal and Navigation Company. The first railroads developed in the early 1830s as feeder lines that connected the mines to the canals. Horses pulled the first freight. By 1833, steam locomotives were beginning to be used on a regular basis on some lines. Freight terminals developed in the region's earliest urban centers like Pottsville, Mauch Chunk, and Carbondale. Philadelphia developed as the primary port for export to New York and New England (O'Bannon et al. 1997, 14; Powell 1980, 7; Taylor 1951, 41).

Consumers were demanding higher quality coal without soil and slate. The development of the breaker answered this call. Joseph Battin patented the basic breaker design in 1843, and the first models appeared in the anthracite region in the mid-1840s. Coal was hauled to the top of the breaker and dumped into a hopper. The hopper released the coal onto a set of screens that separated the coal by size. Conveyed by gravity, a rotating cylindrical mud screen would separate the coal dust and soil from the rest of the coal. Then the coal passed through the breaker rolls, which were cylinders studded with teeth, and broke the coal into smaller pieces. These pieces fell onto screens and finally into storage bins. The coal then passed down chutes where slate pickers manually removed any bits of slate or other impurities (Wallace 1987, 15–16).

Coal mining developed into a major industry by the 1850s and 1860s because of the growing demand to heat houses in the Northeast. Coal also fueled industries, which helped to catalyze the Industrial Revolution (Richards 2002, 7). With the growing demand for coal and the ability to increase production and bring to market a better product, there became a need for more workers. In the 1850s, the Irish fleeing starvation and tyranny ably filled this need. They were a new people, bringing with them new customs and cultural practices. Some observed a "sharp distinction between the 'turbulent' Irish miners and their more respectable 'right thinking' English and Welsh counterparts" (Palladino 2006, 54). "'The Irish,' according to Benjamin Banner, editor of the *Miners' Journal,* demonstrated a 'tendency to retain a distinct, exclusive, nationality'—a tendency eschewed by the 'quieter and better class of emigrants'" (quoted in Palladino 2006, 54).

By the 1880s and 1890s Slavic and Italian immigrants, the newcomers to the region, began to outnumber their predecessors in many of the anthracite communities. The new immigrant increased in numbers from about 2,000 in 1880 to more than 89,000 in 1900. In the same era, those whose families originated in northern and western Europe remained about the same at around 100,000. The growing population of southern and eastern

Europeans created a new nativist sentiment and xenophobic fears among the Anglo-Saxon residents, who are often referred to in the contemporary literature as the "English speakers" (Turner 1977, 10). The coal operators were aware that these new foreign laborers were escaping persecution and tyranny in their native lands. Owners imported more workers than was needed, thus creating a labor surplus. The workers were perceived as interchangeable and disposable, and those in the large unemployed ranks could easily replace them. The coal barons had leverage to keep the wage scale low and their profits high. *The Yale Review* reported in 1898 that "It is also because of a definite and conscious purpose among employers to have on hand a full supply of cheapest available labor. 'Whenever I have had a serious strike, one of my best weapons has been to get in foreign workmen,' was the frank admission of one of the mine owners" (Brooks 1898, 306).

In the nineteenth-century United States, the new industrialists, including the coal operators of northeastern Pennsylvania, were seen as a new type of hero. They were developing new sources of energy by inventing and investing in new machinery. Operators helped to open new lands for cultivation, and they were perceived as bringing modern civilization and order to ignorant, lawless, hungry, and diseased millions. The new capitalist was seen as a special breed of people who were transforming nature into something that was productive and modern (Wallace 1987, 229). Throughout the anthracite region, ceremonial dinners honored the accomplishments of inventors, engineers, and capitalists. Interestingly, the anthracite coal city of Pottsville contains an iron-clad statue standing on a pedestal of the southern statesman Henry Clay. Erected in 1855, it stands sixty-five feet high and can be seen for several miles. While Clay believed that the United States should be self-sufficient in agriculture, he became a hero to northern industrialists because he supported the idea that the country should also be independent in manufacturing. In the face of southern opposition, he advocated for a protective tariff that helped the Pennsylvania iron industry, which in turn supported the coal industry. The anthracite coal town of Ashland is named after his estate. There is also a mine named after Clay (Wallace 1987, 243–44).

By World War I the anthracite coal industry in northeastern Pennsylvania employed about 180,000 workers, the majority living in small patch towns near the coal operations. The living conditions in these patch towns were substandard when compared to the average American household. Many of the coal towns consisted of planned, symmetrically built company houses. The typical patch town had the mine bosses and supervisors living in larger houses, usually on higher grounds. Downhill, or at a lower elevation, sat the duplexes, which the local community called double houses. The company store and the school were usually centered in the community. On the periphery of the planned core settlement were shanty enclaves, small communities constructed in a rather unplanned fashion by the new immigrants with the goal of keeping dry and staying out of the weather. Sanitation and health conditions were substandard at best.

While men's pay was miniscule, women, when not caring for children and gardens, often took on odd jobs to help meet family expenses. Beginning in the late nineteenth century, some of the larger coal towns attracted silk factories and cloth mills employing women at very low wages. These mill owners were escaping organized labor in the urban Northeast and Middle Atlantic states. Despite living in these poor conditions, the new immigrants' plea for better living and working conditions were often easily dismissed by capital. Collective action by the workers was often portrayed in the American press as being associated with radical anarchists. Benjamin Bannan, editor of the *Miners' Journal* and an advocate of the coal operators' free-labor ideology, wrote in 1865 that there was little difference between organized labor and organized crime, and there was little difference between a strike and a riot (Palladino 2006, 131). Bannan frequently targeted ethnicity rather than inequality as a major cause for labor unrest, often blaming the new immigrant for any labor conflict. Bannan and others often linked the rise of labor unions to the stereotype of the Irish miners' "clannishness" and "intemperance." However, these critiques often overlooked the fact that the "respectable" Welsh, English, and native-born miners also worked to organize (Palladino, 2006, 169).

Mining Conditions

In the nineteenth century, the British had developed some of the best engineering practices for mining. However, in the interest of saving time and money, many colliery operations in northeastern Pennsylvania ignored these best engineering practices for the sake of making a quick profit. Many of the mines in the anthracite region had poor ventilation, were inadequately timbered to stabilize the ceilings, and were prone to explosions, roof falls, fires, and floods. Hundreds of miners were crippled and many more killed each year because coal barons ignored these advances in mining safety. They disregarded geologists' reports and continued to use cheap technology at the cost of thousands of human lives. "The monotonous day-to-day toll of lives went largely unnoticed by the public, but the total was grim evidence of the human cost of mining coal" (Miller and Sharpless 1985, 108).

Wallace (1987, 6) notes that the coal operators bemused themselves with illusions. "They told themselves and the world that the absence of high enough tariff on British iron was keeping coal prices too low to make a profit and those careless miners and Irish revolutionaries were largely responsible for the high rate of disasters. These illusions helped the owners and operators . . . to keep on producing coal to fuel the Industrial Revolution" (Wallace 1987, 6).

Whenever an accident found its way into the court system, English and American judges and juries often ruled in favor of the industrial entrepreneurs. The "fellow servant" rule asserted that an injured employee could not hold an employer responsible if a fellow employee caused the accident. It also asserted that the employee was free to leave his master's employment if he considered the conditions unsafe. If a worker continued his employment with a negligent "fellow servant," the worker assumed the risk of the job. The rule would eventually include the risk imposed by an employer who did not provide a reasonable, safe workplace. Because of the precedents in early court cases, issues such as contributory negligence, the "fellow servant" rule, and assumption of risk virtually nullified the employer's risk and responsibility to his workers (Wallace 1987, 279–80).

Coal operators blamed careless miners for the high rate of accidents and deaths rather than their own reluctance to install expensive safety measures. Between 1869, when the Commonwealth of Pennsylvania first recorded the statistics, and 1900, a total of 10,116 mineworkers died in coal-mining operations. In 1870, there were between five and six deaths per one thousand workers, and by 1900 the number had declined to about three deaths per one thousand mine workers (Dublin and Licht 2005, 24). Roof falls and blasting accounted for the majority of the casualties. Benevolent societies tried to provide relief for miners and their families. However, the rate of casualties outpaced the resources of many of these groups. When a worker entered a mine in the morning, he did not know if he would see another day. Peter Roberts, writing in 1901, described the conditions of living in a coal patch town. "Those who, unfamiliar with mining communities, visit them for the first time, generally observe the large number of maimed among the inhabitants. Persons having lost an arm or a leg, or bearing on their face and hands blue scars, or with impaired eye-sight or total blindness, are familiar scenes in mining regions" (Roberts 1901, 152). Roberts then went on to explain that more than 28,000 people had been injured, and about 12,000 died in work-related incidents in the anthracite coal region in the preceding seventy-five years. A short epitaph on one of the tombstones of a former mineworker sums up the experience of the anthracite coal miner. "Forty years I worked with a pick and drill, down in the mines against my will: The coal king's slave but now it's passed, thanks be to God I am free at last" (quoted in Richards 2002, 9).

Coal operators in England developed efficient systems of mine ventilation by the 1860s, which resulted in a significantly lower accident rate. However, ventilation safety measures were ignored in the United States because of the cost of implementation. Poor ventilation often allowed for fire damp, a flammable gas that accumulates in pockets, to collect, which resulted in explosions (Wallace 1987, 201–15). Poor ventilation and an explosion from fire damp caused the Avondale tragedy, which occurred in Luzerne County on September 8, 1869. The mine used an underground furnace and fans for ventilation. The colliery closed for a strike, and when the furnace at the bottom of the shaft was relit, it flashed out of control. Timbers lining the shaft caught fire and it spread to the breaker. The wooden breaker collapsed into the shaft, trapping the entire workforce. The sole airshaft was blocked. About 110 men died of either suffocation or smoke inhalation (Miller and Sharpless 1985, 108; Wallace 1987, 296–99).

A public outcry over the death of these men encouraged Pennsylvania legislators to amend the 1869 measure for mine safety in what became known as the Ventilation Act of 1870. The new law required all mines to have at least two shafts or outlets and prohibited the construction of a breaker over the shaft in furnace-ventilated mines. The law also created the Pennsylvania Office of Inspector of Mines. The office was required to inspect all collieries, record and investigate all accidents, and submit an annual report to the governor (Wallace 1987, 295). Despite these and subsequent safety laws, the coal operator was responsible for appointing safety managers, who were often from the colliery's superintendent's staff, thereby creating the potential for conflict of interest. A significant number of fatalities continued every year (Wallace 1987, 305–11).

In 1885, legislation provided for more-rigid enforcement of safety laws. Additional legislation led to the development of state-funded miners' hospitals, and in 1901, state legislation required the creation of medical rooms in each anthracite mine to provide urgent care (Miller and Sharpless 1985, 113; Trachtenburg 1942, 96). Mines tended to be unsafe and there was a constant threat of collapsing ceilings and fires. Through the 1890s, Prudential Insurance Company considered the mining positions of foreman, runner, driver, door boy, and helper to be among the twenty-five most dangerous jobs in the world (Blatz 1994, 29, 32). In addition, miners breathed coal dust, and over the long-term, if miners survived the brutal working conditions and were able to labor for several decades, suffered from a debilitating lung disease now known as coal workers' pneumoconiosis (CWP), known as black lung disease or simply black lung.

The Response to Poor Working and Living Conditions

In response to these poor working conditions, coal workers made many attempts over several generations to organize. For instance, 1842 is the first recorded major strike in the Schuylkill region. However, it failed, and the miners went back to work without gaining any concessions. The 1848 strike from May 2 to 21 was led by the Bates' Union. We know little about the union, except that it had limited reach and only a few members. Welsh mine workers organized it, however it disbanded by 1850. The significance of this union is that it united mineworkers of all trades and skill levels. Contemporary unions tended to organize by trades. The unionization of all workers associated with mining set the tone and strategy for organizing in the future (Keil and Keil 2015, 48).

Facing famine and oppression in Ireland, the Irish began a large-scale migration to the United States beginning in the mid-1840s. Like many of the newcomers who followed, Irish immigrants were assigned the most dangerous jobs, and were often harassed by the established community of English and German miners. The Irish turned to trade unions to help with their grievances. Local unions and organizations emerged in the 1840s; however, they were not effective in negotiating workers' grievances. Miller and Sharpless (1985, 149–51) explain that Irish settlers faced heightened anti-Catholic sentiment and therefore became more militant in their actions. They, along with other ethnic groups, united behind the first successful regional union—the Workingman's Benevolent Association (WBA) (Rayback 1966, 12).

John Siney, an Irish immigrant miner from Schuylkill County, led the WBA. It was the first union to unite workers from all of the anthracite production fields—north, middle, and south. The WBA was also the first major union to unite along industrial rather than trade-craft lines. At its height, it represented 80 percent of anthracite workers. In 1869 the WBA forced Schuylkill and Lehigh field operators to accept a sliding wage scale tied to the price of coal. The miners agreed to link wages to a sliding scale of $3.00 per ton—known as the "Gowen Compromise." Under this agreement, the miners ended up taking a wage cut as the price of coal dropped. In November 1870, the WBA and the Schuylkill anthracite board of trade agreed to a wage of $2.50 per ton (Keil and Keil 2015, 30).

The WBA also gained legislative support that allowed workers to form and join organizations for their "benefit and protection." Prior to this legislation, workers who organized were subject to prosecution for criminal conspiracy (Trachtenburg 1942, 18–19). However, regionalism and ethnic divides plagued the union. Larger companies, like Franklin B. Gowen's Philadelphia and Reading Railroad, accused the WBA of being dominated by anarchists. Gowen also accused the WBA of being controlled by the Molly Maguires, an activist Irish secret society. By 1875 the WBA changed its name to the Miners' and Laborers' Benevolent Association (MLBA)—identifying itself as an industrial union rather than a trade union. In 1875, Gowen was able to persuade most of the operators to abandon the sliding wage scale tied to the price of coal, and operators agreed to a 20 percent wage cut for miners and a 10 percent cut for laborers. The MLBA responded in what became known as the Long Strike, which lasted for six months (Wallace 1987, 332–334).

The Philadelphia and Reading Railroad refused to negotiate with the union, and Pinkerton detectives infiltrated the MLBA. The Coal and Iron Police attacked strikers and terrorized communities. Miners fought back by damaging and destroying company property. "During this time, the miners exhausted their personal resources. They consumed their small gardens' entire produce and killed off small animals such as chickens, ducks, geese, rabbits, and pigs. The mineworkers and their families were reduced to foraging and hunting to supply themselves with food. After six months on strike, the miners returned to work, beaten down by the companies, and accepted the wage cuts. The miners and their families were literally starving and suffering from severe malnutrition" (Keil and Keil 2015, 34; also see Wallace 1987, 421). The miners' defeat in the strike also meant the end of the MLBA.

The Knights of Labor developed in 1869 as a secret society in Philadelphia and went public in 1878. In 1883 the Amalgamated Association of Miners of the United States, which represented workers from several bituminous coal–producing states, began to organize in the anthracite region. In 1885 the new union, the Miners' and Laborers' Amalgamated Association, consisting of mostly English, German, and Welsh workers, formed a joint committee with the Knights of Labor. The unions called for a strike in August 1887. Both enjoyed some successes. However, they failed to unite the entire anthracite region. Divisions existed on regional and ethnic differences, and these organizations could not overcome these obstacles (Greene 1968, Turner 1977, 12; O'Bannon et al. 1997, 57).

The coal operators kept the mines open throughout the strike and thought they could import Italian and Slavic workers to break it. However, to their surprise, the new immigrants supported the strike and even attacked Hungarian scabs, or "black legs." The strike lasted until mid-March 1888, and the union was defeated. The miners lost to the powerful coal cartel, and both union organizations left the coalfields. By June workers received a 10 percent cut, and independent operators instituted a "dockage confession," which was a waiver against illegal weighing practices. Foremen were also instructed to fire workers they perceived to be militants (Miller and Sharpless 1998, 216–20; Greene 1968; Aurand 1971, 216).

After the strike, a U.S. congressional hearing probed many of the coal operators about the causes of the strike. Eckley Coxe, coal operator from the middle anthracite fields, testified before Congress and told the committee he refused to negotiate with the union. Coxe viewed workers as a force of nature. Labor strife, he said, "has been common to all people who employ others to work for them. I suppose, since the time of Adam." He made analogies to workers as being part of a larger family. "Coxe operators and men," he said, "are very much like a man and his wife; they quarrel and fuss but they have got to live together." Later, he admitted that when the first time Coxe workers struck, "it almost made me sick. [But] it is like a man when his first child had a tooth, he thinks it is dreadful, but when his third child has the fifteenth tooth, he does not think so much of it" (Coxe quoted in Holt 2001, 12). He believed that a remedy for strikes lay in the development of common schools, so that labor and capital could better understand each other (Holt 2001, 12).

While much of the investment in anthracite extraction came from outside interests, the middle coalfield had several powerful independent coal operators who owned mining interests, railroads, land trusts, banks, lumber operations, local mills, and ironworks, as well as controlling politics, police, businessmen, and clergy. One account described them as, "lords of a small fiefdom they had made through their own enterprise" (Miller and Sharpless 1985, 225). These independent coal operators saw themselves as guardians of their community. They set the pattern for labor relations and refused to deal with unions (O'Bannon et al. 1997, 34).

The New Workforce

While the mid-nineteenth-century coal industry workforce mainly consisted of "English speakers," labor upheaval and strikes later in the century encouraged coal operators to recruit a new, foreign workforce. They reasoned that if laborers spoke different languages it would be more difficult for the workers to communicate and organize. Beginning in the 1880s, in the midst of continued labor disagreements, coal operators began to recruit workers from eastern and southern Europe, with the goal of ethnically dividing the labor movement, which was dominated by the Irish (Miller and Sharpless 1998, 170).

Future immigrants facing various hardships in southern and eastern Europe were easily enticed to work for the coal operators. Generally known as the "Slavs," they came from eastern Europe and the Balkans, and included Poles, Ukrainians, Czechs, Slovaks, Serbians, and Croatians, fleeing Czarist Russia and Austria-Hungary. Non-Slavic immigrants came from Lithuania, Austria-Hungary, and Italy. Many acted on an opportunity to escape their deteriorating working and living situations. For instance, in Russia and the Austro-Hungarian Empire, there was a growing aristocracy that misused agricultural lands. Smaller farmers and peasants overworked their lands and depleted the soils. Populations doubled in the second half of the nineteenth century in some regions of Europe, including what is now Poland. By 1900, 50,000 people died annually owing to starvation in the Galicia region of Poland. People also emigrated because they were escaping a repressive feudal system. They were also fleeing political and religious repression (Greene 1968, 25–26; Miller and Sharpless 1998, 172–73; O'Bannon et al.1997, 59; Stolarik 1980). Therefore, it proved easy to entice many who were facing difficult times to accept offers to work in the anthracite region.

Coal operators saw the new immigrants from southern and eastern Europe as "cheap men," and employing them at lower wages would enhance their profit margins. Companies contracted with labor agents in Europe to recruit new workers. The new immigrants were willing to work longer hours than English-speaking miners. They also accepted dangerous work conditions and substandard living arrangements, conditions that the English-speaking miners did not tolerate. The new immigrants also worked for lower wages (Warne 1904, 77). Many unskilled workers who were making a few cents an hour in Europe were enticed by the pay of a dollar a day, or more for unskilled labor in the coalfields (Miller and Sharpless 1998, 173).

There are stories in the anthracite coal region about the poor treatment new immigrants faced when they arrived at the patch town. They were dropped off at the railroad station and from that point on had to fend for themselves. Some initially found shelter in the mines and they foraged for food. Others scrambled to create makeshift housing out of scavenged lumber from the colliery (Korson 1938, 125; Miller and Sharpless 1998, 176).

In 1880, 90 percent of the foreign-born residents in northeastern Pennsylvania were "English speakers." However, by 1900, the number of foreign-born "English speakers" dropped to about 50 percent. During this same period, the numbers of Slavic descent grew from 2 percent to more than 40 percent. About one-half of them, or 34,000 Slavs, worked in the mines (Miller and Sharpless 1998, 181; Roberts 1970 [1904], 20). On the local and national level, the "English speakers" felt increasingly threatened by the large influx of immigrants, and resorted to scientific racism to maintain and justify their sense of superiority and continued exploitation of the newcomers.

Racialization of the New Immigrant

The creation of race and the racialization of the other is one of the most haunting episodes in human history, and still holds power on our everyday relations. The concepts of race and ethnic identity are charged with meaning, and they develop in different ways. Orser (2007, 8) explains that ethnicity is created from the inside, while race is imposed from the outside based on perceived biophysical differences as well as cultural practices, religious beliefs, traditions, and a combination of physical and cultural attributes. Therefore, racialization is the process of assigning people to groups based on physical or cultural characteristics, which helps create the perception of inferior or socially unequal groups. Racialization creates racially meaningful groups that previously did not exist. Those classified as "other" are seen as inferior to the group creating these classifications (Omi and Winant 1983, 51; Orser 2003, 2007, 9).

Race thus emerged as a social classification by which to create "otherness" in much the same way that the concept of savagery developed. Theodore Allan (1994) and many others (for example, Breene and Innes 1980; Handlin and Handlin 1950; Jordan 1978, 278; Kulikoff 1986; Morgan 1975) discuss the invention of the white race, using the Chesapeake region to examine the development of racism in the New World. Until the 1660s statutes applied to enslaved blacks were similar to those for white indentured servants. Legislators applied stiff penalties to blacks and whites equally for running away, drunkenness, and carrying arms. Over time, these regulations grew less stringent for white servants, but little changed for blacks (Handlin and Handlin 1950, 244; Morgan 1975). Differential treatments of blacks and whites developed as early as 1662 and were codified in provincial laws. Many American scholars identify Bacon's Rebellion in 1676 as the major catalyst for developing the modern concept of race and the creation of whiteness in America. Bacon's Rebellion was a revolt by indentured servants, poor freedmen, and enslaved people against the propertied classes. Though the rebellion was eventually suppressed, colonial administrators subsequently codified laws that divided the laboring classes along clear (and artificial) racial lines. Laws were created that privileged white skin, and black slaves were increasingly subordinated with subsequent legislation. There was an alliance between classes based on biophysical traits. Faye Harrison (1998, 621) explains that it was a "cross-class alliance—premised on the payment of a psychological wage" (Harrison 1998, 621).

Racialization continued through additional legislation during the eighteenth and nineteenth centuries (Shackel 2003). The new racial ideology that developed in western Europe and the Americas structured political, social, and economic inequalities as countries from these regions conquered and exploited a large part of the world. This new racial ideology contradicted the development of Renaissance ideals that promoted freedom, democracy, equality, and human rights. Europeans naturalized their new racial attitudes by focusing on physical differences and concluded that Africans and Indians were, by the laws of nature or God, a lesser form of human beings (Smedley 1998, 694). Race identity took priority over all other forms of identity that had once guided socialization and group interaction.

When Irish Catholics fled the potato famine in the 1840s, they were greeted with nativist sentiments, and political organization like the Know Nothings began their anti-immigration campaigns in the industrial Northeast (Baker and Patterson 1994, 2). The racism the Irish were exposed to had its origins in the British Isles, where the Irish were subjugated by the English and identified as in opposition to the English as "Irish savage" (Orser 2003; Sacks 1994). The categories of race being constructed created social hierarchies and barriers that justified and legitimized existing or emerging relations and practices in the United States.

By the 1870s and 1880s, the Irish began to claim some political power, and many challenged their nonwhite status (Sudarkasa 1968, 158). The Immigration Restriction League (IRL), founded in 1894, emerged as one of the prime organizations shaping public opinion about immigration. Three Harvard graduates, who were among the Boston Brahmin elite, founded it. Henry Cabot Lodge, a U.S. senator from Massachusetts, served as one of the IRL's prominent members. At first, they were concerned about the growing political power of the Irish. Subsequently, their mission and constituency became much broader, opposing the new immigration (Solomon 1956). The organization developed a campaign against the "Slav, Latin, and Asiatic races" (Solomon 1956; Sudarkasa 1968, 158).

The momentum and power of the Immigration Restriction League persuaded President Theodore Roosevelt in 1907 to create a congressional commission to investigate immigration. In 1911, the U.S. Immigration Commission completed its work—a forty-two-volume report, *The Reports of the Immigration Commission,* also known as the Dillingham Report. The reports described the distribution of immigrants throughout the United States and reported on education, crime, employment, insanity, prostitution, and so forth (Pavalko 1980, 60).

Much of this work relied on the scientific racism scholarship popular at the time to create racial categories. In particular, the work of Daniel Britton, a Philadelphia-trained physician who became professor of ethnology and archaeology at the Academy of Natural Sciences and the University of Pennsylvania, had a significant impact in the scholarship found in the *Reports of the Immigration Commission*. Britton described five races based on a combination of physical and cultural traits. They were white Eurafricans, black Austroafricans, brownish Asians, coppery Americans, and dark Insular peoples. Britton (1890a, 56, 350; 1890b, 47–50, 96–107) used classifications of races that were already being commonly used in political discourse and in newspapers for several decades. He explained that biophysical characteristics differentiated these ethnic groups. These different groups were races and they were part of a natural hierarchy. He explained that there was a clear hierarchical order and that the white race was superior. Britton also created white-skinned buffer races that separated Anglo-Saxon groups from the other darker-skinned subspecies (Patterson and Spencer 1994, 23).

Turn of the twentieth century racial classifications divided the Slavic population into three groups according to territories. The eastern Slavs include "the Great Russians, the White Russians, and the Little Russians, or Ruthenians." In the western portion of the Slavic territory "are found the Poles, the Bohemians, the Moravians, the Slovaks and the Serbs. On the southern portion of the territory are found the Serbians or Serbs, Croatians, Bosnians, Montenegrins, Slavonians, Dalmatians, Slovenes, Bulgarians, and Macedonians" (Roberts 1970 [1904], 27–28). The *Dictionary of Races or Peoples* deviated from Britton's five racial categories and further divided the Italians into "Celtic" and "Iberic" groups, following Sergi (and hence Niceforo), separating the group into north and south Italians (D'Agostino 2002, 331). The southern Italian was described as a long-headed, dark, Mediterranean race of short stature. They were closely related to the Iberians of Spain and the Berbers of northern Africa. The southern Italian may have some infusion of blood from Africa (U.S. Senate 1911, v. 5, 3, 8). They were described as "vivid in imagination, affable, and benevolent, but excitable, superstitious, and revengeful" (U.S. Senate 1911, v. 5, 127). The northern Italian was described as, "cool, deliberate, patient, practical, and . . . capable of great progress in the political and social organization of modern civilization" (U.S. Senate 1911, v. 5, 82). Those from the southern part of Italy faced harsh living and working conditions, and the largest influx to northeastern Pennsylvania came from regions south of Rome. In *The Dictionary of Races or Peoples* the Slav was identified as "carelessness as to the business virtues of punctuality and often honest, periods of besotted drunkenness among the peasantry, unexpected cruelty and ferocity in a generally placid and kindhearted individual" (U.S. Senate 1911, v. 5, 129).

The Dillingham Commission reinforced the racial divide by distinguishing between the "old" and "new" immigrants, with the inference that the "old" are more civilized than the "new."

> In what follows, the old and new immigration will be considered to include the following races or peoples: Old—Dutch and Flemish, English, French, German, Irish, Scandinavian, Scotch, and Welsh. New—Armenian, Bohemian and Moravian, Bulgarian, Ser[b]ian and Montenegrin, Croatian and Slovenian, Dalmatian, Bosnian and Herzegovinian, Finnish, Greek, Hebrew, North Italian, South Italian, Lithuanian, Magyar, Polish, Portuguese, R[o]manian, Russian, Ruthenian, Slovak, Spanish, Syrian, and Turkish. (U.S. Senate 1911, v. 1, 22)

Racialization became reinforced through additional legislation, in the case of the southern and eastern European, often in the form of quotas on immigration and in some cases, taxation (Shackel and Roller 2012; Shackel 2016). The new racial ideology that developed in western Europe and the Americas structured political, social, and economic inequalities. They naturalized their new racial attitudes by focusing on physical differences and cultural differences, often concluding that the "other" was a lesser form of human being (Smedley 1998, 694). Those controlling the narrative relegated the new immigrant as low status and they constantly portrayed them as "culturally backward, primitive, intellectually stunted, prone to violence, morally corrupt, undeserving of the benefits of civilization, insensitive to the finer arts" (Smedley 1998, 695).

The racialization of the different groups created differences, and these differences were actualized in relationship to whiteness. Whiteness is based on the social construction of race and class (Bodkin 2000, Fine 2004). Whiteness is linked to relations of power and domination. The direction of power between white people and others is naturalized and becomes embedded in the fabric of American culture (DiAngelo 2011, 56; Feagin 2006). "Whiteness is the assumed norm, and when non-western European immigrant populations come to the United States they are usually ethnically and racially identified as 'other' or not white, based on biophysical traits, or subscribing to different norms and customs. These traits signify the creation of the 'other,' and emphasize the non-white characteristics, such as hair, skin color, eyes, stature, etc." (Low 2009, 81).

Much like the Irish were racialized, the immigrants coming from southern and eastern Europe were faced with the same social Darwinian classification where, in some cases, they were seen as uncivilized and only partially human. For instance, in 1885, in a report of the Iowa Bureau of Labor Statistics, a laborer expressed his racial opinions in terms of whiteness. He explained,

"The Bohemians . . . will get a job in preference to a white man" (quoted in Pavalko 1980, 61). In a 1911 hearing before the House Committee on Immigration and Naturalization, which was part of the data collection by the Immigation Commission, U.S. Senator John L. Burnett of Alabama stated that, "the southern Italian is not a 'white man,' nor is the Syrian" (quoted in Pavalko 1980, 61).

Peter Roberts (1970 [1904]) wrote an early-twentieth-century description of the Slavs in northeastern Pennsylvania. His book *Anthracite Coal Communities* is a racialized treatise that describes the new immigrants in a derogatory framework, and it follows much of the racist thinking of the era. Roberts describes the newest immigrants to the region, the Slavs, as "his intellectual capacity clogged 'by the weight of the centuries' is not equal to that of the German and the British workman. . . . The Russians . . . are a few centuries behind the rest of the civilized world" (Roberts 1970 [1904], 25). The author then states that the Slavs show promise of social progress and assimilation. Roberts notes, "The Sclav is a good machine in the hands of competent directors. He is obedient and amenable to discipline, courageous and willing to work, prodigal of his physical strength and capable of great physical endurance. . . . He thinks slowly and is willing to follow the lead of others, but when the Sclav is once set in motion in a given course, he is there to stay" (Roberts 1970 [1904, 36).

Roberts (1970 [1904]) continues to cement his racialization of the Slavs by declaring that all the Slavs drink. He explains that they look upon lager much in the same way that the English view tea drinking and the French look upon wine. Though they observe the Sabbath, "they buy drink, dance, sing ribald songs, play cards, etc., on Sunday without scruple" (Roberts 1970 [1904], 53). A Baptist deacon who visited a Slav enclave, Mahanoy City, reported, "It was terrible; saloons full blast; singing and dancing and drinking everywhere; it was Sodom and Gomorrah revived; the judgment of God, sir, will fall upon us" (Roberts 1970 [1904], 53). Reinforcing his racialization of the Slav, he continued to write about work in the anthracite mines in terms of whiteness. "The seams changed, there was more brawn needed, more powder per ton of coal must be burned, and 'white men' refused to work under conditions that meant more labor, more expense, and less pay. Then they called in the willing Slav, the submissive Lithuanian, and the work was done" (Roberts 1912, 55).

The racialization of groups based on places of origin allowed for the *Reports of the Immigration Commission* (U.S. Senate 1911, v, 16) to develop a scheme that related job competencies to race in the mining industry.

Table 1.1. Race preferred by mining operators for specific occupations.

Pick miners	Fire bosses	Engineers	Foremen	Stablemen, Blacksmiths, etc.
American	American	German	English	German
English	English	American	American	American
Welsh	Scotch	English	German	English
Irish		Irish	Irish	Welsh
German				Irish
French				Slovak
Lithuanian				
Polish				
Slovak				
Ruthenian				
Belgian				
Italian, North				
Italian, South				

Source: *Reports of the Immigration Commission* (US Senate 1911 v 16)

Those most preferred by the coal operators were of northern and western European ancestry, while Slavs filled the ranks of the lower third and the southern Italians were the least preferred.

Creating a racialized view of the new immigrant allowed for the easy exploitation of the newcomers. While the "English speakers" received relatively good compensation for their work, by the middle of the nineteenth century the coal industry began to boom and new foreign workers, predominantly Irish, came to the anthracite fields, only to be treated poorly by the capitalists in power. The same treatment occurred with the new immigrants of the late nineteenth century from southern and eastern Europe. Because of their placement on the evolutionary hierarchy scale, coal operators felt justified not promoting them to the better jobs and not paying them an equal wage or supplying them with domestic accommodations equal to those found outside the region. The living conditions in the shantytown only reinforced the stereotypes imposed upon the new immigrants. Many of the newcomers were familiar with unions and the ferment of revolution for the working people in their native lands. Their exploitation only led to similar ferment in their new homeland. The process of strip mining developed at the turn of the twentieth century as heavy machinery became available. However, the amount of anthracite coal extracted by strip mining did not exceed the coal extracted from the underground until 1961.

CHAPTER 2

The Lattimer Strike/Incident/Massacre

Nationalists and Unions

In the 1890s, mineworkers began to support other organizations that would work to counter the strength of the coal-mine operators. The United Mine Workers of America (UMWA), founded in 1890 in Columbus, Ohio, had some initial success in organizing in the anthracite region. They began by recruiting members of different ethnic groups, and each local retained their ethnic identity (Turner 1977, 12–13). The UMWA established its first local in the anthracite region (in Shamokin) in 1892. The union had a stronghold in the Schuylkill (southern field) region. Much like the previously failed efforts to organize, without the support from the other fields, and without the support of the newest and fastest growing population, the eastern and southern European workers, the union's presence in the anthracite region quickly diminished (Miller and Sharpless 1998, 220; Aurand 1971; Roy 1907).

The American Protective Association, a nativist organization that was anti-immigrant and anti-Catholic, worked to halt the Slavic and Italian immigration into Pennsylvania by promoting restrictionist legislation. In 1889, they supported successful legislation that required miners to have two years of experience underground and pass an exam that was written in English, before they could be certified as a miner. Though the bill passed easily, enforcement was spotty at best and had little effect. Many non–"English speaking" mine workers hired interpreters to take the exam for them (Greene 1968, 115–17). By the mid-1890s, many coal operators and politicians realized that the law worked against the recruitment of workers who were of Anglo-Saxon background. Instead of waiting the two years to qualify for the exam to earn higher wages, those of western European descent went to other parts of the state and country to earn a decent wage (Turner 1977, 14). John Fahy, district president for the UMWA, complained that the penalties for not enforcing the act were too weak. He supported a revision of the act that included stiffer penalties, and the prohibition of interpreters for the exam, thus making it difficult for non–"English speakers" to succeed (Turner 1977, 14).

The coal operators were strategically enticing many more workers to the region than they could keep employed. Fearing the new immigrant, as well as feeling the competition for jobs, the Pennsylvania legislation passed a resolution in 1895 and 1897 to petition Congress to enact legislation restricting immigration. The resolution described the new immigrant as a "growing evil highly injurious to American workingmen and dangerous to American institutions" (*Laws of the General Assembly* 1895, 643; 1897, 546). John Fahy began working for discriminatory laws that reinforced the strength of the mineworkers who were U.S. citizens, those of English, German, Welsh, and Irish descent. He believed, along with the other miners, the new Slav and Italian immigrants were willing to work for lower wages, and therefore kept wages low. With the backing of his constituents, Fahy lobbied the state legislature for the passing of the Campbell Act, also referred to at the Alien Tax law, a company tax of three cents per day for each alien worker. The mine owners opposed the act. However, when passed and signed into law, the coal operators passed along the tax to their employees, believing they would not have any trouble with their immigrant workers. After all, the operators believed they were unintelligent and docile, and they would probably not resist (Blatz 2002, 43, Greene 1968, 127). This tax compounded the effects of the new immigrant earnings, which were 10–15 percent lower wages than their Anglo-Saxon counterpart for the same duties. The country was struggling to get out of a depression and the coal market was depressed. The men were only working an average of two to three days per week (Miller and Sharpless 1998, 222; Novak 1978). The Campbell Act went into effect on August 21, 1897. In July 1897, Fahy wrote in the *United Mine Workers Journal* that he not only approved of the tax, he also thought it should have been higher (Greene 1968, 127). "What a world of good this law would do to the American citizens who try to earn their living in the coal mines if the tax were one dollar per day. I have an idea it would also do the foreign-born . . . a power of good by keeping them out of the coal mines where all is cruel poverty and misery" (Fahy quoted in Blatz 1994, 54).

Beginning of the Hazleton-Area Strike

Strikes were common in coal country, sometimes lasting a day or two, and other times lasting weeks or months. Labor unrest in the Hazelton area in August 1897 was part of a chain reaction of events that led to one of the most significant confrontations between labor and capital. On August 12, 1897, at the Honeybrook Colliery near McAdoo, Gomer Jones, division superintendent for the Lehigh and Wilkes-Barre Coal Company, posted a new work rule that required the twenty mule drivers to stable their animals in a central place. The new location meant that the mule boys would be required to work two extra hours a day without compensation (*Hazleton Sentinel*, August 16, 1897; Greene 1968, 130; Turner 1977, 23; 1990, 2). Angered by this new arrangement, the mule boys went on strike the following day. They lined the road to the stables to monitor and ensure that the mules did not return to the mines. When Jones heard about this development, he went to meet the boys with a club in hand. An argument erupted and when the boys refused to return to work, Jones struck one of them on the head with an ax handle. Jones reasoned that work discipline was lacking and he was trying to restore order (Novak 1978) (figure 2.1).

Incensed by Jones's brutal treatment of the mule boys, about 350 miners armed with guns and clubs persuaded each of the Lehigh and Wilkes-

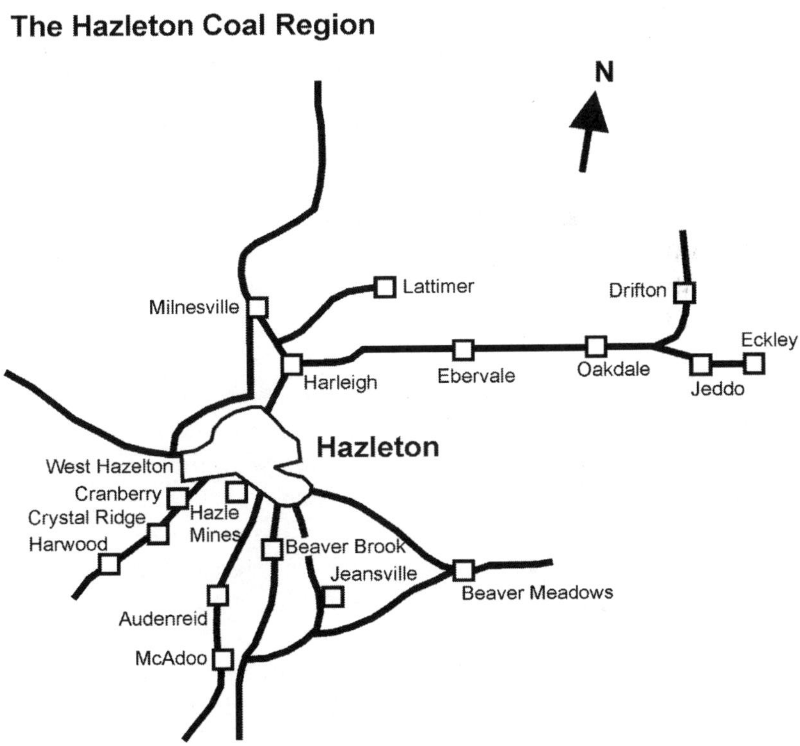

Figure 2.1. Map showing Hazleton and some of the surrounding coal-patch towns. (Image by Sara Downard.)

Barre collieries to close. By the end of the day more than three thousand workers joined the cause. Seeing this unrest develop, John Fahy took this opportunity to once again try to organize the anthracite miners. He was uncertain about his reception because he and the UMWA supported the Campbell Act. Surprisingly, the strikers greeted him with "cheer after cheer" (*Daily Standard,* August 19, 1897). By the end of August 19, the workers in the south towns of Hazleton voted overwhelmingly to join the union. Fahy had enlisted more than eight hundred members, organized into six chapters representing six different ethnic groups (*Daily Standard,* August 19, 1897; Miller and Sharpless 1998, 223; Novak 1978). At the same time, a Slavic and Anglo-Saxon committee traveled to New York to meet with company officials, and reported back to the miners. The workers received some concessions in pay as well as their demand to use their own doctor. The miners agreed to return to work (Blatz 2002; *Daily Standard,* August 19, 1897; Greene 1968, 131; Novak 1978; Turner 2002).

However, when the miners received their pay on August 21, for the first time they noticed the three-cent per day tax deduction. Within a week, the tax was wide-felt. On August 26 slate pickers walked off the job, and within a few days several thousand miners marched to the various collieries and shut them down. When the Lehigh and Wilkes-Barre Coal Company did not take any action against Gomer Jones, the workers from this company joined the strike (*Daily Standard,* August 26, 1897; Miller and Sharpless 1998, 223; Novak 1978).

The new immigrants took the lead on the strike. All of the workers demanded a 15 percent wage increase, an end to the company store, and the right to pick their own doctor. The immigrants also wanted the same pay scale as their American counterparts. For instance, at the time of the strike, the English-speaking miner earned $1.93–$1.97 while the "Hungarian" miner received $1.65 (*Hazleton Sentinel,* August 27, 1897; Aurand 1971; Miller and Sharpless 1998, 224; Novak 1978). The workers were also seeking a uniform wage scale in the region. Wages for the same jobs differed by as little as a few cents to $1.00 (*Hazleton Sentinel,* September 2, 1897b). By the beginning of September, more than five thousand striking mineworkers had closed many of the coal operations in the Hazleton area (*Daily Standard,* August 28, 1897; Turner 1990, 10; Greene 1968, 131).

Within several weeks the United States District Court declared the Campbell Act unconstitutional. While several companies promised that they would refund the taxes already withheld from the miners' paychecks, Superintendent Drake of the Pardee collieries, which included the ones in Harwood and Lattimer, declared that the company would continue to withhold the tax from its employees until there was a decision by the U.S. Supreme Court (*Hazleton Sentinel,* August 31, 1897a). This attitude infuriated the workers.

The miners marched to each of the Lehigh and Wilkes-Barre collieries and successfully closed them down. Lehigh and Wilkes-Barre workers formally voted to strike on September 1. Marches were organized throughout the region, closing other collieries (Blatz 2002; *Hazleton Sentinel,* September 2, 1897a; *Hazleton Sentinel,* September 2, 1897b; Novak 1977; Turner 2002). The *Hazleton Sentinel* (September 2, 1897) headline subtitle read, "The Foreigners Refuse to Listen to Counsel." By September 3, 1897, the strikers had closed all of the mines of the Lehigh and Wilkes-Barre Coal Company (Pinkowski 1950, 8). The *Philadelphia Inquirer* (September 3, 1897) headline read, "Mob Rule up in Luzerne District." The *Wilkes-Barre Times* offered, "'Twas a grand stroke to march eleven miles and close up four breakers. Napoleon's greatest achievements were overshadowed. And as the setting sun cast its last ray over the distant mountain the grand army of striking Huns, Italians and Slavs marched to their homes to enjoy the calm and quiet peace after a day of war" (quoted in Greene 1968, 136).

Though many of the "English speaking" miners were willing to return to work, the new immigrants significantly outnumbered them, and the strike continued (*Hazleton Sentinel,* September 4, 1897). The new solidarity shown by the southern and eastern European workers meant that future relations with labor could change. Distressed over the continuing strike, the coal operators met in Hazleton to discuss their concerns. They feared that workers would begin sabotaging the coal operations, and the Coxe Company and Calvin Pardee and Company purchased Winchester rifles to arm staff and deputies (*Daily Standard,* August 31, 1897; Novak 1978, 31–34).

Demonstrations continued over the Labor Day weekend. Concerned about the potential destruction of property, the coal company superintendents ordered Sheriff Martin to return from his vacation on the New Jersey shore and organize a posse. George Turner (1977, 28) points out that it was not a public official who summoned him to Hazleton—rather, it was Superintendent Lathrop of the Lehigh Valley Coal Company. The coal operators wanted their property protected and they wanted to break the strike. Martin arrived at Wilkes-Barre on Monday, pledged to defend the property of the coal operators, and agreed to deputize one thousand men, if necessary (Turner 1990, 10).

The strikers had closed down most of the mines on the south side of Hazelton. The coal-mining operators made it clear to Martin that they would hold him responsible for damages or work stoppages to their operations. The sheriff declared a state of disorder, which allowed him to form a *posse comitatus*. By Labor Day, September 6, he had eighty volunteers, mostly consisting of men with English, Irish, and German backgrounds with some connections to the coal industry. He armed them with Winchester rifles and metal-piercing bullets (plain bullets of heavy bore) and no. 8 buckshot. The Coxe Company ordered ten thousand rounds of ammunition (Pinkowski 1950, 10). Martin, along with the sheriffs in the surrounding counties of Carbon and Schuylkill, issued proclamations banning mass gatherings and demonstrations. Defying this order, miners marched from colliery to colliery to get other miners to join them, and in effect, shut down many of the mines (*Daily Standard*, September 7, 1897a; *Daily Standard*, September 7, 1897a).

On Labor Day, between three thousand and four thousand men of many different nationalities marched together in the south towns to commemorate the day. It was the first time so many men had marched under the banner of the UMWA in this region (*Daily Standard*, September 7, 1897). After Labor Day, the demonstrations continued and the miners struggled to systematically close all of the collieries in the Hazleton area (Miller and Sharpless 1998, 230). For the next several days, Martin and his deputies traveled almost nonstop trying to prevent the strikers from closing collieries. Though often blocked by the deputies, the number of immigrants on strike increased each day (Turner 1977, 29).

The March to Lattimer

The Harwood local of the UMWA held a meeting the same day in a schoolhouse. Through translation, John Fahy was able to organize the workers at Harwood at a cost of 25 cents per worker. Many miners presented their demands, which included a ten cent per day wage increase, a reduction in the price of powder at the company store, and not paying for a company doctor out of their wages. The miners at Harwood, who worked for Calvin Pardee and Company, readily joined the union and continued the strike. While the Harwood miners were on strike, the Lattimer miners, also employed by Calvin Pardee and Company, were still working. News came to Harwood that the miners at Lattimer would join the strike if the strikers made it to the colliery. In order for Pardee to consider their demands, it would be necessary to close the entire company (Pinkowski 1950, 11). Joe Mahalko, president of the Harwood local of the UMWA, cautioned the strikers to march without weapons, to keep the peace, and to destroy no property (Turner 1977, 30).

On September 10, at about noon, protestors started their march in Harwood and set off for Lattimer with about 250 men and two American flags (figure 2.2). The demonstration slowly grew in numbers (Turner 1977, 31). They met with the sheriff and his posse outside West Hazelton on a public highway. The sheriff took out a piece of paper from his pocket and told the marchers that it was against the law to march to Lattimer. He never read the proclamation. A fight erupted and Ario Pardee Platt confiscated one of the American flags and ripped it to pieces (*Hazleton Sentinel,* September 10, 1897b; Miller and Sharpless 1998, 231; Novak 1978, 122). Police arrested two of the miners, and several more were injured, one suffering a broken arm. The police chief of Hazelton said that marchers could proceed, if they agreed to travel around the city limits. They agreed and moved peacefully and unarmed. The strikers believed they had the American right to free assembly and freedom of speech. One of the deputies was heard saying, "Let them go until we get to Lattimer, and then we will shoot them down" *(New York Times,* February 19, 1898; *The Press,* February 10, 1898) (figure 2.3).

Sheriff Martin and his posse took a trolley to Harleigh, where they waited for the approaching marchers. By the time they reached the outskirts of

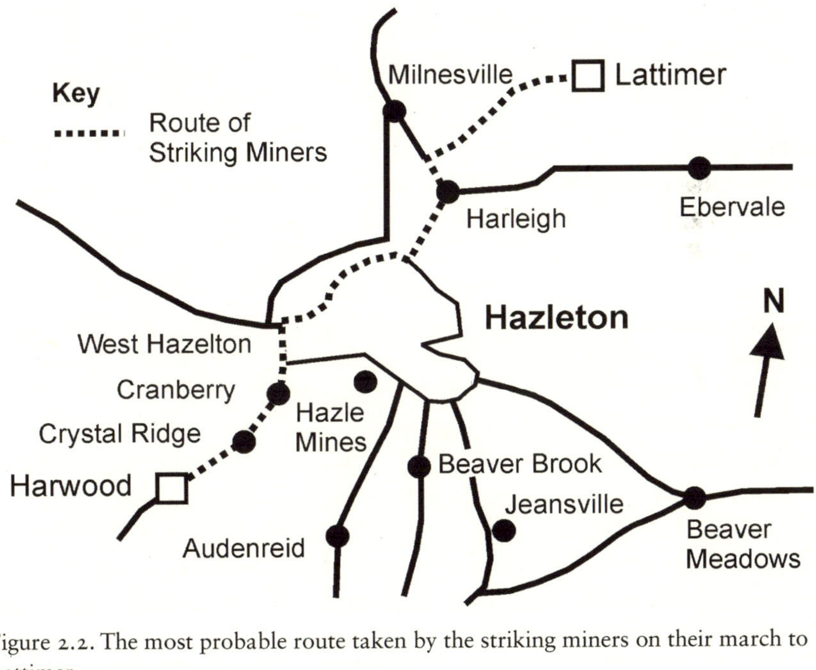

Figure 2.2. The most probable route taken by the striking miners on their march to Lattimer.

Figure 2.3. Coal miners on strike and marching toward Lattimer, September 10, 1897. (Courtesy Pennsylvania State Archives, MG 273, Charles H. Burg Collection.)

Lattimer, the strikers grew to four hundred men. The mineworkers stopped when they saw the sheriff and his deputies stationed on top of a hill in front of Farley's Hotel. After pausing and conferring with each other, they then proceeded up the hill toward breaker no. 1. When the strikers turned right on Lattimer Road, the sheriff moved his men to a point ahead of the miners and before the first house in Lattimer (Pinkowski 1950, 13). Additional coal and iron police joined the posse at Lattimer, and Martin had about 150 men to protect the colliery. Edward Pinkowski (1950, 13) wrote: "The sheriff and his Slav-haters jumped off the car and three companies formed a horseshoe across the public highway in front of the first house. The colliery whistle rallied more deputies who were stationed at No. 1 and No. 3 breakers. The road curved from the trolley bank on the ridge of the hill and cut the space between the first house and the trolley tracks practically in half."

Sheriff Martin waited under a tree, which later became known as the "Massacre Tree." The deputies lined up along the front-yard fence of the westernmost house in the patch town. The men were positioned on the north side of the road so that they could easily encounter the long axis of the marchers. On the south side of the road stood an embankment for a trolley that was several feet high. Unarmed, the strikers marched into this indefensible position and met Sheriff Martin at the tree. Martin had his pistol drawn (*Hazleton Sentinel,* September 10, 1897).

Apparently, the posse was ready to shoot, as several trolley passengers later reported they overheard the men talk about how many strikers they could kill. The social Darwinist ideology prevalent during the time made it easier to justify the thought of killing these men as though they were game, because they were seen as uncivilized and, to some, not even human. The whistle blew, which signaled the shutdown of the colliery. Company police joined the posse and the sheriff. At 3:45 P.M., with an American flag in hand, about four hundred strikers reached the road outside Lattimer Mines. The sheriff met the strikers about forty to fifty yards in front of the line of deputies (Turner 1977, 32). From newspaper accounts and oral histories, Ed Pinkowski (1950, 14) recalls the confrontation between the miners and the strikers in his monograph.

> Sheriff Martin, clutching the proclamation, walked forward to meet the oncoming men who were holding the American flag.
>
> "Halt!" the sheriff commanded. "Where are you going?"
>
> "We go to Lattimer mines," several voices replied.
>
> "Stop," barked the sheriff, "you are disobeying the law. Go back."
>
> "Go on," someone shouted, "Go on."
>
> "Who said, 'go on'?" snapped the sheriff, grabbing the man nearest him by the coat collar. He lowered his right hand and drew his revolver. . . . Anthony Novantny, who said he was an American citizen, pushed through the ranks to explain to the county official—
>
> "We are not disobeying the law," he said. "We have no clubs; we are not going to kill or murder. We want to go on through the town, as is the privilege accorded any man, and not to interfere with anybody." (Pinkowski 1950, 14)

Suddenly there was pressure from the rear ranks, and the sheriff pulled a striker from the second row to the side of the road. Martin attempted to seize the American flag; however, he was met with resistance. The workers moved forward. A scuffle broke out and Martin drew his pistol, then pulled the trigger; however, it did not fire. Several witnesses claimed that the sheriff then shouted "Fire!" and "Give two or three shots," although Martin would later deny it. The flag-bearer was hit and mortally wounded. The men, realizing the shots were not blanks, began to run away from the scene. Witnesses standing nearby reported several men running for cover at a nearby schoolhouse and being shot in the back. The Lattimer schoolteacher explained, "When the strikers got in front of the deputies I heard one shot. Then some of the men stopped, and some continued down the road. Then there was another shot and a succession of shots and the strikers began to run up towards the schoolhouse. . . . When the volley was fired then men began to fall, killed and injured." They made it about one hundred yards, or about halfway to the schoolhouse (*Plain Speaker,* February 5, 1898). To shoot these men, the deputies had to climb up the trolley berm to take aim (figure 2.4).

Others fleeing in other directions, as far as 150 yards away, were shot in the back. The shooting ended after two minutes (*Daily Standard,* September 11, 1897; *Hazleton Sentinel,* September 10, 1897; Pinkowski 1950, 14). The

Figure 2.4. "Firing on the Miners," *Philadelphia Inquirer,* September 12, 1897.

Daily Standard (September 11, 1897) described the actions of the deputies as "taking cool, deliberate aim, bringing down their men as if some choice fowl." The *Hazleton Sentinel* wrote, "The groans of the dying and screams of the wounded filled the air" (*Hazleton Sentinel,* September 10, 1897). Edward Pinkowski (1950, 14) describes, "Those in front dropped like wheat stalks before a scythe. Steve Urich, a Slovak immigrant who was carrying the American flag, fell into the ditch near the sewer pipe that ran beneath the street railway. 'My God,' he cried in Slovonian, 'that is enough.' He was, according to eye-witnesses, the first man killed."

Deputies came out of their ranks, followed the strikers for about thirty yards, and shot miners trying to reach the thin rows of trees on the other side of the trolley bank. The *Daily Standard* (September 11, 1897) declared, "It was not a battle, because the strikers were not aggressive, nor were they on the defensive, because they had no weapons of any kind and were simply shot down like so many worthless objects, each of the licensed life takers trying to out-do the others in this butchery."

The *Hazleton Sentinel* (September 10, 1897) reported, "A battlefield was reproduced and many a battle has been fought with less carnage." Nineteen men were dead and about another thirty-eight were wounded. Only in five cases did death result from front fire. All the rest came from direct or indirect rear fire, a result of being shot while fleeing (*Daily Leader* quoted in Palmer 1913, 115). The following day, the governor sent the state militia to Hazelton. Thousands of people from the region came to view the massacre site and to participate in the funeral ceremonies (Greene 1968, 140). Funerals held in the different churches drew as many as eight thousand mourners.

Immediately after the massacre, the *Baltimore Sun* (September 11, 1897) reported that Sheriff Martin admitted the marchers were on a public road, going to Lattimer and acting in a peaceable manner. Sheriff Martin also gave conflicting reports about how the shooting started at Lattimer. In his initial statement Martin explained, "I halted the marching column and read the proclamation, but they refused to pay attention and started to resume their march," he told the *Philadelphia North American.* "I called the leader to stop but he ignored my order and I attempted to arrest him. I hated to give the command to shoot and was awful sorry that I was compelled to do so, but I was here to do my duty." Later that day, after meeting with his attorney, and in all subsequent interviews, Martin explained that he never gave the order to fire (Pinkowski 1950).

Whether Martin gave the order to shoot or not is still debated when discussing the history of the Lattimer massacre. Nevertheless, the shooting of the protesting workers did happen and the event created significant debate and discussion about the role of organized labor and the racialization of foreigners in turn-of-the-century America. How the victims were treated in the community over the short and long term set the stage for how they were memorialized for generations to come.

The Strike Lingers

Clergy leaders, as well as John Fahy of the UMWA, spoke the day after the massacre in Hazelton at a gathering of two thousand citizens who came out to show their support for the miners. He, along with other community leaders, stressed the importance of being peaceful. Fahy urged the workers to unionize. The gathering unanimously adopted a resolution condemning the murders of the miners and stating that the sheriff and the posse should be prosecuted to the fullest extent of the law (Turner 1990, 20).

After the burial of the massacre victims, the UMWA rallied five thousand workers to enroll in the union. Pardee employees consisted of about 2,700 of them. By Monday, September 13, the strikers of Lehigh and Wilkes-Barre Coal Company settled on an agreement and went to work (*Daily Standard,* September 13, 1897). On Tuesday (September 14) hundreds of Pardee mine workers gathered at the company store to receive an answer about their demands. Soldiers of the Thirteenth Regiment stood nearby. Augustus W. Drake, the superintendent of the Lattimer mines, came out of the store and met the men. He said "no" to all of their demands. Drake explained that they had nothing to gain by being idle and he asked them to go back to work. The men refused (Pinkowski 1950, 27). About eleven thousand men remained on strike (Greene 1964, 209–10).

Two days later (September 16) many of the men began to return to work when they heard that the Campbell Act was suspended (Greene 1968, 144–45; 209–10). Women throughout the region thwarted miners' attempts to return to work. They were described as "a body of brave hearted women . . . mobilized near the eastern portion of town and accomplished more than five times as many men could have. [They] drove out a band of Italians form the Honey Brook strippings, and made them flee like so many rats . . ." (*Daily Standard,* September 17, 1897). During the evening of September 19, "Hungarians" marched through the streets of Lattimer beating cans and kettles, notifying the men to continue the strike (*Santa Fe Daily New Mexican,* September 20, 1897).

While men began to return to work in Lattimer on September 20, Mary Septak, also known as Big Mary, rallied about one hundred Slavic and Italian women to support the strike. She operated a boarding house in Lattimer and knew the hardship stories of the coal miners, about their working conditions and their near starvation wages. One of her sons worked in the Lattimer breakers before he died. She had buried nine of her ten children, who died from various illnesses (Pinkowski 1950).

Septak assembled the women, some with babies on their hips, and stopped a number of men from going back to work. Known as "The Amazons" by the National Guard and newspaper reporters, they spent a few days keeping the mines closed at several collieries by shouting at the returning workers, knocking tools from their hands, and pelting them with rocks. The newspaper headlines read, "Women Force Men to Quit," and "Women Cause Slight Trouble at Harwood," and "Another Raid by Amazons." Miners stood at a distance cheering on the women, and some newspapers reported that men dressed as women made up part of the Amazon group. In another account, the women stood on a culm bank, stoning the strikebreakers, and men and boys waited in reserve behind the bank in case they were needed (*Daily Standard,* September 17, 1897; *Hazleton Sentinel,* September 20, 1897; Learn-Andes 2004).

As a result of these developments, the National Guard came to Lattimer to help reopen the colliery. They formed two lines, fixed their bayonets, and clicked their triggers for effect. The women sparred for about forty minutes by hurling insults at the Guard. Septak and others climbed the culm bank and rallied in defiance. Septak "raised her little sword aloft, cried out a defiance and at the same time stood on her tip toes and whirled about several times, elevating her petticoats and exposing a pair of stout ankles" (*Hazleton Sentinel,* September 20, 1897). As the Guard slowly advanced, she "bowed to the forces in a rather meek fashion and slowly moved down the railroad." The *Wilkes-Barre Record* criticized the women for not meeting the role expectations of Victorian femininity. It noted, "The appearance of women as a factor in a coal region strike is a novelty of a not very pleasing nature" and condemned the men for sanctioning their role (quoted in Greene 1968, 143–44; Pinkowski 1950, 27).

Jay Hambidge (1898, 825) described Septak as, "the most forcible and picturesque character in all the mining region. In her peculiar way she is a queen, and rules things with a high hand" (Hambidge 1898, 825). He described her as the most troublesome of all of the foreigners during the events surrounding the Lattimer massacre. "No professional agitator had half the force for mischief that this woman exerted" (Hambidge 1898, 825).

After the National Guard subdued Septak and her followers, the strike began to wane as many of the companies reached an agreement with the workers.* The Lehigh and Wilkes-Barre Coal Company agreed to a 10 percent wage increase and six to eight cents more per carload of coal. The Lehigh Valley Coal Company and Calvin Pardee and Company had not reached an agreement with its workers until the end of the month (*Daily Standard,* September 21, 1897; *Hazleton Sentinel,* September 20, 1897; Turner 1977, 43, fn 173).

National Reaction

Local and national newspapers provide an overview of the varying emotions and sympathies related to the bloodshed. For instance, the headline in Hazelton's *Daily Standard* (September 11, 1897) read, "Yesterday's Butchery—A Mob of Heartless Deputies Fire into a Throng of Marchers and Accomplish Deadly Work." The *Pottsville Republican* (September 11, 1897), which tended to be pro-capital, had a headline, "Strikers Shot in Cold Blood—Horrible Scene of Carnage Near Hazelton Yesterday." The incident made it to major national papers in New York, Detroit, Washington, D.C., Boston, St. Louis, and Pittsburgh (Turner 2002, 15). The *Hazleton Sentinel* (September11, 1897) wrote, "The fact that the victims are exclusively foreigners has detracted, perhaps from the general expression." However, Reverend J. V. Moylan of St. Gabriel Roman Catholic Church described the incident as a "brutal and unjustifiable massacre." Rev. Francis I. Pribyl, a former priest of St. Joseph's Slavonian Church in Hazelton, noted that, "The slaughter of them in cold blood is the most high handed piece of butchery that has ever been perpetrated upon a peaceable people and a sad commentary upon the boasted freedom of fair Columbia, whose extended arms we are taught to believe are continually outstretched to the down trodden and oppressed people of other lands" (*New York Journal and Advertiser* quoted in Turner 1984, 128).

The editor of the Slavic language newspaper, *Amerikánsko-Slovenské Noviny* (Pittshburgh) declared; "The mountainsides of Hazleton are drenched with Slovak blood, and pitiable orphans and widows, fathers and mothers, brothers and friends raise eyes brimming with bitter tears to heaven pleading, O God, is there justice in this life?" (Turner 1990, 18). *Amerikánsko-Sloven-*

* One of the great mysteries of Lattimer is what happened to Mary Septak. She does not appear on the 1900 U.S. Federal Census. Some in the region have speculated that she may have changed her name to avoid reprisal, and others believe that she moved away after the coal company blacklisted her. An article in the *Plain Speaker* (November 11, 1899) notes that she died in her house while doing chores. The article referred to her as "Big Mary" and recalled her heroic stand against the National Guard.

ské Noviny was referring to the well-known high mortality rates associated with mining.

The *Hazleton Sentinel* (September 15, 1897) provided a review of selective newspapers across the country. They all contain a racist view about the strikers and condemn them, rather than the sheriff and his posse. For instance, the *New York Times* wrote that the sheriff's actions probably stopped an episode of widespread killing and looting. The *Washington Star* described the strikers as "ignorant men" and "a dangerous class" who do not know the language and customs of the land, although the coal operators should be responsible for bringing them to the United States. The *Pittsburgh Gazette* described the strikers as "vicious fellows." The *New York Sun* described the protesters as "an ignorant and turbulent crowd." The editor criticized how some papers have made heroes of the deceased. These papers, the editor explained, are "inciting the ignorant to violence, and thus luring them on to death." The *Philadelphia Inquirer* called these men part of a foreign swarm and said that the United States should create laws to prevent the migration of immigrants if they did not know how to write and speak the language.

A Polish newspaper published in Scranton rephrased Lincoln's Gettysburg Address. For those who died in Lattimer, "May their death not be in vain, may they become the patron saints of the working people in America" (*Stranz,* September 18, 1897 quoted in Turner 2002, 22). The headline in the Slovak newspaper in Pittsburgh, *Amerikánsko-Slovenské Noviny,* read, "Massacre of Slavs—in the Freest Country Under the Sun—People are Shot at like Dogs—Slavs are the Victims of American Savagery"(quoted in Turner 2002, 22; Stolarik 2002, 35). Associating the Americans with the term "savagery" reversed the definition of civilization, where westerners were civilized, and other races were not. The Slovak newspaper is questioning the accepted hierarchical order.

While the men returned to work, various forms of sabotage in the collieries occurred, including machine breaking and even the burning of breakers, which were timber frame construction. The mysterious burning and destruction of the Beaver Meadows breakers occurred during the evening, after the miners had gone home (*Hazleton Sentinel,* September 22, 1897). James Scott's (1985) examination of everyday tactics of peasant resistance serves as a model for understanding power and power relations in this mining community. Resistance in a capitalist world could take on several forms, including malingering and sabotage (see for instance Juravich, 1985; Paynter 1989, 386; Scott 1985; Shackel 1996, 59–61). This class conflict is quite evident in the anthracite coalfield region and it continued whenever there was a conflict between labor and capital.

Racism and the New Immigrant in Lattimer

In the spring of 1897, just prior to the Lattimer massacre, the governor of Pennsylvania appointed a committee to survey the bituminous and anthracite regions and to report on the living and working conditions of the workers and their families. The W. B. Meredith Committee visited several communities and interviewed 311 witnesses representing businessmen and directors of relief organizations, as well as many others associated with the mining in the region (Turner 1990, 6). In 1896, the average number of days of employment in the Hazelton region was 178 days, or 57 percent of full employment—for a year based on a six-day workweek, for fifty-two weeks a year (or 312 days). These conditions persisted into 1897, which meant that miners were working an average of three days a week and living in starvation conditions, a situation which many of the new immigrants thought they left when they came to America (Turner 1990, 7). The report claimed that conditions in the mines had been deplorable for a long time. The report continued:

> The men in and about the collieries have been employed not more than two or three three-fourths days per week, earning on an average $4 per week, upon which, in many instances, they were compelled to support large families, in some cases as high as eleven members, paying house rent and coal and the necessities of life, which to this committee seems an impossibility. In mining centers there is at all times more or less destitution and want in many instances occasioned by profligacy. Matters throughout the anthracite region have been gradually growing worse, and never in the history of this section of our State has the condition of the men employed in and about the mines been so deplorable or of such long duration. (Legislative Investigation 1897 in Turner 1990, 7–8)

These deplorable conditions were becoming more public with government investigations, and finally the massacre at Lattimer led to other inquiries. However, advocacy for the miners did not occur in the anthracite region without derogatory, racist comments describing the foreign-born workers. In 1898, *The Century* magazine sent several reporters to the anthracite region to describe everyday life of the miners and their families. Two of the correspondents, Henry Edward Rood and Jay Hambidge, focused on the conditions in Lattimer.

In Rood's (1898) article, "A Pennsylvania Colliery Village: A Polyglot Community," he used the "English speaking" mining families as a foil to compare the customs and culture of the new immigrants from southern and eastern Europe. Rood made a comparison to the earliest miners in the region from Wales and England. He claimed they had an abundance of work and a family could make a good wage. While they paid higher prices at the company store, they also received relatively good wages. "Quickly they established themselves and ceased to speak regretfully of childhood associations in Wales or in England" (Rood 1898, 809).

He justified the social and economic inequities that existed in the coal-patch town (as well as in the United States) and stated that the new immigrants were uncivilized and did not want or care to have a higher standard of living until English-speaking "labor agitators" encouraged them to strike and riot. He noted that the miners should not complain about their wages because they now earned about four to five times the amount of what they were earning in Europe. However, even while earning a lower wage than their "English speaking" counterparts, they did tend to save money. He implied that the "English speakers" tended not to be as thrifty. The reason, according to Rood, is that the "English speakers" tended to clothe and educate their children and they took pride in the houses they occupied. Sunday dress was important to the English-speaking miners, and they liked to take their families to dances and picnics. They bore the same expenses as skilled laborers elsewhere (Rood 1898, 815).

Rood continued his description of the Lattimer residents and wrote that the Italian workers were about three centuries behind the American standard of living—based on their clothing, food, shelter, and general intelligence. Because of their position in the social and evolutionary hierarchy, Rood (1898, 818) explained that many of the new immigrants were content with their low wages and huddling in their shanties. The union agitators made them want to strike. They ate "half spoiled vegetables and fruits that could not be sold to the English-speaking people. . . . They have their own churches and amusements and weekly newspaper" (Rood 1898, 818). They would have never thought of engaging in a general strike, and the events at Lattimer would have never occurred if not for the "English-speaking labor agitator." Rood (1898, 818) continued his racist rant: "The ignorant hulking Slovaks and Polacks, and the brawny cunning Italians, who formed the mobs, would not have thought of raiding through the lower end of Luzerne County had it not been for the politicians and agitators." Justifying the deaths at Lattimer, Rood (1898, 818) then explained that the only way that this riotous mob could be stopped was from a volley from the Winchester rifles.

Rood (1898, 812) described the evening activities in the shantytowns. Many sat around and told stories about the day's work and relayed stories they heard in the mines. "Some of the more intelligent" talked about the American government and discussed the importance of gaining citizenship so that they could vote and ensure that the "white men" were no longer in the majority. Rood continued to paint a picture of the new immigrants from eastern and southern Europe as being uncivilized and murderers. He noted that others lived in the patch town plot in the nearby woods, and discussed how to get rid of some of the hated enemy, like a member of the coal and iron police force—or a priest who antagonized them (figure 2.5).

In his racist rant, Rood (1898, 811) described the new Slovak immigrants as "decreased in the scale of civilization until those who have come to the anthracite fields during the first half of the present decade are, as a rule, much more dangerous to the body politic than the excluded Chinese; for not only are they eager to work for wages on which an 'English speaking' family would starve, but they are superstitious and murderous, and do not hesitate to use dynamite if they desire to blow up one of whom they particularly hate." Rood (1898, 818), expressing the xenophobic fears of the time, explained that the new cheap labor was taking work away from the American, German, Scandinavian, and British workmen. He expressed a fear of an influx of anarchists. He called for a restriction of immigrants, although few politicians wanted to create quotas because they felt immi-

Figure 2.5. Italian section of Lattimer with children on the road. Image taken soon after the Lattimer massacre. (Courtesy Pennsylvania State Archives, MG-273, Charles H. Burg Collection.)

grants could be induced to join their political organization. However, he recognized and saw as important the work of the Immigration Restriction League. He wrote that he feared the new immigrants from eastern and southern Europe would have the same political influence as any American as soon as they were naturalized. Rood explained that the new immigrants cared nothing for our country, and that they were a people who firmly believe in witchcraft. He then compared them to half-domesticated animals (Rood 1898, 820–21).

Complementing Rood's work in Lattimer is Jay Hambidge's (1898) "An Artist's Impression of the Colliery Region," published in the same edition as Rood's piece. Hambidge provided commentary and sketches of individuals in Lattimer. He noted some young boys walking through town and described then as "intelligent-looking younkers." He also noticed a girl of about six years old hauling water from the settlement well in front of the church. Her load would be enough for someone three times her age. Hambidge (1898, 825) then explains, "This is the first evidence—of which we will see more later—of the woman as the animal, the chattel, the thing to be possessed for its usefulness, as a piece of furniture, a cow, or a mule. A little later in life (not much later—six years is enough) this mite of a child will be, not married, but given and taken in marriage; and the certificate from the priest will be to her husband as a bill of sale—documenting evidence of possession."

Hambidge (1898, 826) describes the boarding house that Mary Septak operated along with her husband in the shanty area of the patch. In the living room were seven beds and eight trunks, where about a dozen men resided. Each bed had pictures of the crucifixion, the Virgin Mary, and patron saints. At one end of the room were two men sleeping in a bed with their clothes on. They worked the night shift.

A stream of sulfur-water pumped from the mines ran in a ditch through part of the town. Barren lands existed around the culm piles as chemicals leached into the ground and killed grasses, trees, and brush. At the edge of the shanty patch were scattered stockade enclosures, which served as gardens for the mining families. The ground was rocky, and it was a challenge to grow anything in this soil. "To the immigrant just arriving from Italy, the colliery town must seem a realization of desolation itself" (Rood 1898, 809). Describing the environment around the shanty patch, Hambidge (1898, 828) wrote, "Toward the north, a fertile valley reaches forth, and is bounded only on the horizon. It is one of the most beautiful and productive valleys in all the great State of Pennsylvania. Behind us, despair, ignorance, strife, and struggle for mere existence; before us, the beautiful valley seems a land of infinite promise."

Hambidge (1898, 822) explained that "truly this is the land where life is held cheap." He explained that while visiting the office of Calvin Pardee and Company, close to the no. 1 coal breaker, he heard an explosion. About two hundred yards away a premature explosion killed a man. His entire chest was ripped away. There were no other comments in the coal company office, and the incident appeared to be very matter-of-fact to the supervisors. Later that day, someone mentioned the incident and asked if the man was dead. "I think he is," was the reply, "for I saw his little girl coming from school crying" (Hambidge 1898, 822). Accidents were very common in the coalfields, and there were frequent deaths from accidents, such as the one just mentioned, as well as frequent losses of arms, legs, or hands, and injuries to eyesight and hearing (Hambidge 1898, 822).

The Century magazine articles by Hambidge and Rood are racist in tone, and they reinforce the ideals of racialization that existed in the anthracite region as well as throughout the United States. The new immigrants often inhabited the poorest sections of town, usually shanties with poorly constructed houses and no sanitation. They were not seen as white, and they were often described as lacking any form of modern civilization. They spoke different languages, wore different clothes, and had practiced different customs. These attitudes toward the new immigrants set the stage for one of the great miscarriages of justice in American labor history.

CHAPTER 3

A Great Miscarriage of Justice and the Growth of the UMWA

The Trial

Despite the tendency of the dominant culture to see the victims of the Lattimer massacre as uncivilized, there was tremendous local, national, and international pressure for justice. The community established the National Prosecuting and Charity Committee of the Lattimer Victims (NPCCLV) with Father Aust of St. Stanislaus Church as its president. Four of its thirteen members were clergymen. The mission of NPCCLV was to raise money to prosecute the sheriff and his deputies and to help support the families of the victims (*Daily Standard*, February 28, 1898; Turner 1984, 134).

The Austro-Hungarian government took an interest in the case and on September 30, 1897, the Austrian Ambassador to the United States, Baron Henglemiller, came to Philadelphia to meet with his consul. Two weeks later they dispatched the consul's secretary, D. Theodorovich, to investigate the Lattimer incident. Spending a week in the Hazelton area, Theodorovich found ten of the nineteen killed and eleven of the thirty-eight wounded were Austrian and Hungarian citizens. A formal diplomatic note to the United States from the Austro-Hungarian government protested the killing and wounding of its nationals (*Daily Standard*, October 1, 1897, October 15, 1897; Turner 2002, 23).

Warrants were served on the sheriff and all of the deputies, and the court set bail at $5,000 for each man for each count of murder and $1,000 for each man for each count of felonious wounding. Calvin Pardee, owner of the Lattimer mines, arranged the bail of $438,000 (*Daily Standard*, September 23, 1897; Novak 1978; Schooley 1977, 71). After the incident a coroner's hearing investigated cause of death. On September 27, 1898, four of the six jurors found that the strikers were mercilessly shot to death. The two jurors who did not concur explained that there was a strong suspicion of unlawful violence (*Grand Forks Daily Herald,* September 28, 1897). Based on this decision, the case of the murder of the Lattimer victims went to trial before a jury.

It took two days to select the jury. None of the jurors were familiar with the work of miners and none were of Slavic descent (Culen 1977, 53; Palmer 1913, 62). The district attorney selected the death of one of the miners, Michael Cheslock (Cheslak, Ceslak), as the first case. He was the most prominent victim claimed by the Austro-Hungarian government. Cheslock was an active member of the community and a church trustee, and only a few months earlier had applied to be a U.S. citizen (Beik 2002, 77; *Plain Speaker,* February 3, 1898).

Unfortunately, no complete records of the trial remain. They have disappeared from the Luzerne County Courthouse, although much of the trial can be analyzed using newspaper accounts. There is also a rather biased account outlined in the autobiography of Henry Palmer (1913), *Fifty Years at the Bar and in Politics,* one of the attorneys for the defendants. The court case transcripts were still available at the time of Palmer's writing of his autobiography and there are selective excerpts of the trial in his book.

The legal issue in the case was whether riotous conditions existed in the Hazelton area because of the strikers' actions. The prosecution argued that the threat of a riot in Hazelton did not warrant the creation of a *posse comitatus,* and therefore its existence was illegal. They also maintained that the miners were legally using the highway to march to Lattimer and that the *posse comitatus* was an illegitimate body that acted in an unjustifiable manner, and therefore committed a criminal act of murder (Turner 2002, 26–27). The miners' attorney argued that the strikers did not threaten the public safety and that they were peacefully using a public highway when they were stopped at Lattimer. Therefore, the killings at Lattimer were a criminal act (Turner 1984, 144; *Plain Speaker,* February 4, 1898).

One of the first witness of the trial, Morris Engleman, was on the road near Lattimer when the strikers approached the deputies. He testified that he did not see the strikers carrying arms. Another witness was Daniel Ferry, a grocer from Hazleton, who said he was driving along the road when he witnessed the shooting. He testified that the strikers were marching peacefully in a quiet and orderly fashion. Ferry said he was about twenty yards away from the deputies at the time of the shooting. He heard a single shot and saw one of the strikers fall. Then he heard a volley of shots, and several more strikers fell. After the volley, there were a number of separate shots fired. Ferry testified that he saw a wounded man lying near the trolley tracks get up. As he was trying to walk away, he was shot down again. He also testified that he saw one of the deputies, William Raught, run after the strikers as far as the trolley track and shoot toward the schoolhouse, where several of the miners were heading for shelter from the onslaught (*Plain Speaker,* February 11, 1898; *Daily Standard,* February 11, 1897; *New York Times,* February 19, 1898).

The jury also heard from a line of witnesses who were injured and maimed during the incident. George Gasperick was carried in by four Slavs and lowered into the jury box because he was paralyzed from gunshot wounds. "I can't see well now," he whispered, "and every time I move my head I feel something rattle there. I am unable to raise my arms or legs" (*Plain Speaker,* February 18, 1898; Pinkowski 1950, 36). John Slabonick came to Wilkes-Barre to testify even though he had a bullet in his head. Andrew Meyer, an experienced attorney for the sheriff, tried to twist his words and state that he was forced by the miners to join the march. He told the defense attorney that it did not require compulsion to fight injustice. Casper "Mike" Dulass testified next. He was wounded with a bullet that went through his back, pierced his lung, and left his body through his breast. During his testimony, he broke out in violent fits of coughing (*Plain Speaker,* February 12, 1898; Pinkowski 1950, 35–36). "Another witness, Bernard Ruman, whose spinal column was grazed by three bullets, enacted a scene I would never dare to see. He lay pinching his fingernails into the woodwork of a stretcher and biting his lips in agony, as he lay flat on his back directly in front of the judge. He groaned as each morsel of testimony came from his parched lips" (Pinkowski 1950, 36). Andrew Hannes and Adam Lapinski showed their scars to the jury. Both were wounded in the back while running away from the deputies.

Another miner, Thomas Parish, testified that he was shot through the back and the bullet crushed through his bone and tissue, and it came out through the front of his body. He was asked to undress and show his wound to the jury, which caused a snicker in the court (*Daily Standard,* February 12, 1898). Michael Srochak told the jury how eight deputies followed the fleeing strikers and shot them as they ran from the scene. "I was standing near the deputies when the firing began. A man fell dead by my side, and I dropped instead of running. After the volley about eight of the deputies followed the strikers for about 30 yards shooting all of the time" (*Daily Standard,*

February 12, 1898; *New York Times,* February 11, 1898). Marin Lacher described the deputies' cruelty in the firing into the crowd. He explained how he saw one of the deputies kick Andrew Maier as the latter lay in the road severely wounded, with his leg completely mangled from gunshot wounds. Michael Julas testified that he heard a deputy give the order to fire. Julas was hit in the back while fleeing the scene. The bullet pierced his lung and came out an inch above his heart (*New York Times,* February 11, 1898). Angelo Matz, a rag picker, needed an interpreter to convey what he saw. He told the jury that after a man was shot and wounded, and lying near a culvert face down, a deputy came up to him and shot him three times in the back (*Daily Standard,* February 18, 1898).

Those wounded in the incident were taken to the State Hospital and examined by Dr. Keller. He testified that he treated forty-three wounded men, and he described the wounds of each. Keller noted that five men were wounded from direct front fire, five from indirect front fire, fifteen from direct side fire, six from indirect side fire, six by direct rear fire, and six from indirect rear fire. Direct fire was described as the full force of the bullet or buckshot. Indirect fire refers to the missile passing through another body or limb, losing energy, and becoming lodged or embedded instead of passing entirely through the body (*The Press,* February 5, 1898).

The defense acknowledged that they did kill the miners. "They kill—a fact which they and all men with hearts in their breasts deplore, but they commit no crimes. They do their duty. They do not violate, but obey the law" (quoted in Palmer 1913, 67). The *Daily Standard* (February 22, 1898) responded with a headline: "'A Charging Mob!' That is What Attorney Ferris Calls the Defenseless Men Who were Shot in the Back at Lattimer." The defense recalled several incidents of violence prior to Lattimer and described the protesters as an angry mob. They reportedly forced people to join their ranks, and those who did not were beaten. Homes were broken into and men were dragged out to join the march, while others fled into nearby woods (Palmer 1913, 68). The defending lawyer told the jury that when they arrived at Lattimer, "they were six to eight hundred strong armed with clubs, iron bolts, bars and revolvers" (Palmer 1913, 70). He described them as a ruthless mob who looted along the way to Lattimer, breaking into houses, and physically coercing men to join the march (*Daily Standard,* February 22, 1898; *Wilkes-Barre Record,* February 4, 1898, quoted in Palmer 1913, 61). Several witnesses testified that the strikers had rocks and clubs, and intimidated miners to join the march (*Daily Standard,* February 25, 1898; *Plain Speaker,* February 23, 1898; *The Press,* February 21, 1898).

On March 8, the prosecuting attorney, Mr. Scarlett, in his closing statement argued that the marchers were unarmed and proceeded peacefully. "Witness after witness told you that the crowd was not riotous. They marched along the highway as labor has a right to march—believing that the flag that they carried would protect them—not the red flag of anarchy, but the flag it is a pride to be wrapped in the death—the same flag that was flying to the breeze and stopped the invading march of a nation. Instead of stopping them again, the deputies formed a line to deliver as enfilading fire" (*Daily Standard,* March 8, 1898).

In his closing argument, defense attorney Lenahan described the strikers as lawless hordes that came from the steppes of Asia and had found their way here. "The history of the Hun and Slav in the old country is that of mischief and destruction. And they marched under the Attila ruthlessly over Europe. No home was left sacred or virgin too pure for their assault." He noted that there were two types of Americans, one that loves the country and the other—the anarchists. He called the flag holder a barbarian, and asserted that this case was against the foreign invaders who came to America to destroy our peace and liberty (*Daily Standard,* March 8, 2016; *Plain Speaker,* March 8, 1898).

The next day, defense attorney Henry W. Palmer (ex–attorney general for the Commonwealth of Pennsylvania) made an additional plea. Palmer explained, "The question of fact to be found by the jury before any one can be convicted is, Who killed Mike Cesla (Cheslock)?" He then replied to his own question: "There is not one syllable of evidence identifying or tending to identify that man. The man who did the deed is not himself conscious of it. How can the jury say that Mike Cesla [Cheslock] came to his death by a gunshot wound inflicted by any one person? It was certainly done by someone, but the identity of the man will never be disclosed until the last great day" (*Plain Speaker,* March 9, 1898). Palmer later argued that the assembly of men marching on the highway was unlawful (*Plain Speaker,* March 9, 1898).

Palmer noted that if the miners won the case, it would be a triumph for anarchism and only strengthen the cause to overthrow the government. Lattimer would become a rallying cry for socialists, anarchists, and haters of organized government (*Daily Standard,* March 8, 1898; *Plain Speaker,* March 8, 1898). Palmer described the prosecutors as villains (*Daily Leader* quoted in Palmer 1913, 119).

> Men will point to Lattimer, he said, as the place where born a movement that would wrest from American people the liberty so dearly bought. In time to come,

> unless strong effort is put forth to check it there will reign in this country anarchy and anarchists, pledged to usurp every honest right. Men, unless throttled now, will rise up in rebellion against our government to tear down the stars and stripes and fix over the highest places the red flag of riot and confusion. Bitter and scathing was Mr. Palmer's denunciation of the barbarian horde who will, if they can, point with pride to Lattimer as the place where was shed the blood of their martyrs (Palmer quoted in *Daily Leader* in Palmer 1913, 120).

Palmer described Fahy and other labor organizers as "low brow-beating villains who are crawling over the country unable to earn an honest living" (Palmer in *Daily Leader* quoted in Palmer 1913, 121). He called Father Aust and other priests who helped the miners "Servants of the Prince of Peace, who have taken vengeance into their own hands. They have associated with themselves a saloonkeeper, whose business it is to send souls of men to hell!" (*The Press,* March 5, 1898).

In closing, Palmer noted that if the jury found the sheriff and his posse guilty for doing their duty, it would be the end to peace and order. It would be the end of the Anglo-Saxon way of life and it would lead to a reign of oppressive military despotism (Palmer quoted in *Daily Ledger* in Palmer 1913, 146; *Plain Speaker,* March 9, 1898; *Daily Standard,* March 9, 1898).

Judge Woodward instructed the court to decide if riotous conditions existed and, if they did, were the sheriff and his deputies justified in their response, and was the response appropriate to these conditions. The jury reached a verdict within a half hour with only one ballot. On the morning of March 9, the defendants were acquitted. The jury noted that the killing and the wounding of the men at Lattimer were justified because riotous conditions existed (*Daily Standard,* March 8, 1898; March 10, 1898).

Responding to reporters, Father Aust explained, "The whitewashing the deputies received today will not hide the fact that there was a grave wrong done at Hazleton. . . . Many of the [deputies] are good men and are innocent of this crime of malicious killing as I am. There are about eleven men among the crowd who I believe are guilty of malicious shooting and if they are not punished it will be an outrage on justice" (*Daily Standard,* March 10, 1898).

The Reaction

While the U.S. Department of State agreed with the trial's outcome, the Austro-Hungarian government protested to the U.S. government about the unfair trial of its citizens in a U.S. court. The Austro-Hungarian government petitioned that they "were guilty of no acts of violence, nor in any opposition to local authority which might justify the measures taken against them. . . . The deceased were fired upon without cause, and were in retreat, and that these acts constitute a violation of the rights of Austrian citizens" (*Daily Standard,* October 15, 1897).

The majority reaction was that of outrage, although there were a number of papers that sided with the deputies. Generally, newspaper accounts throughout the region expressed their doubts about the American judiciary system and questioned the results of the trial. The Slavic-language newspaper *Amerikánsko-Slovenské Noviny* reported on the outcome of the case and stated, "They will kill, they will shoot, and grant you no mourning or tears of relief. And somehow they will make it all come out within the limits of the law" (quoted in Turner 2002, 29).

A *New York Evening Journal* (March 10, 1898, 5) sketch titled "What is Crime in Pennsylvania, Anyhow?" captures the rage of a portion of the United States as the nation was keeping a watchful eye over the court proceedings (figure 3.1). Underneath the sketch, the caption read, "In the mining regions deputy sheriffs have a right to kill unarmed strikers on sight. They may pursue them and shoot them down, as they would mad dogs. An in the end deputy sheriffs are glorified as martyrs—not hanged as murderers."

ANOTHER TRIAL CERTAIN
FOR SHERIFF MARTIN AND THE OTHER
LATTIMER MURDERERS.

Prosecutors Say the County's Honor Demands It.

Jury Feared Coal Barons and Wouldn't Punish "Best Families."

VERDICT IN ADVANCE.

So Also Was Judge's Charge. Sheriff Tells What He'll Do in Next Trouble.

WHAT *IS* CRIME IN PENNSYLVANIA ANYHOW?

It seems to l settled that in the mining regions deputy sheriffs have the right to kill unarmed strikers on sight. They may pursue them and shoot them down as they would mad dogs. And in the end the deputy sheriffs are glorified as martyrs—not hanged as murderers.

Figure 3.1. "What is Crime in Pennsylvania Anyhow?" *New York Evening Journal,* March 10, 1898.

The outrage, however, was not unanimous, and capital had its opinions endorsed in the *New York Sun*. The editorial wrote that Sheriff Martin and his deputies faced a mob at Lattimer, and the verdict was a "victory for law and order" (*New York Sun* from Palmer 1913, 168–69). A Philadelphia newspaper, *The Press* (March 9, 1898), agreed with the verdict. In fact, it stated that there should be a memorial erected to honor Sheriff Martin so that alien populations would understand the meaning of liberty. Though there were graves as a consequence of this incident, this sacrifice was necessary. "The law once more is supreme . . . and Sheriff Martin and his deputies are righteously vindicated. . . ." The editorial explained, "If these men had conducted themselves in the outskirts of Prague or Naples as they did here practically in the outskirts of New York and Philadelphia, they would have been shot down at the end of the first day of their revolt, and not, as was the case, at the end of the first week. And there would have been no trial whatever, not even a Court of inquiry into the conduct of the captain of *gendarmes* responsible for the act. . . ." (*The Press,* March 9 1898). *The Press* describes those who lived at Lattimer as "the ignorant survivors of the Lattimer riot."

The Yale Review commented on the general sentiment of an anonymous coal operator. The author (Brooks 1898, 307) wrote,

> In the enforcement period of idleness which come with the shifting conditions of the market innumerable occasions for troubles like those at Lattimer may at any moment appear. "What can you do with such wild beasts when they get off their heads but shoot 'em?" were the words, which the writer heard, and it may be, in any given moment, that the social safety demands quick, sharp and bloody enforcement of the law? It is, however, a very sinister state of affairs when conditions, which have been definitely encouraged by the mine owners and by our general policy of immigration, have come to be such, that, in their very nature, they are certain to breed chronic outbreaks like this in the Hazleton district. (Brooks 1898, 307)

Assistant Attorney General Henry M. Hoyt reported to the Department of State regarding the incident at Lattimer. He noted that the strikers were all foreign born and of Slavic background. "As a class, they are ignorant, and from our point of view uncivilized. Compared with such people as our own, they represent a status of civilization to find which generally we should have to go back several centuries" (Hoyt quoted in Palmer 1913, 193). Because they were considered foreigners and uncivilized, Hoyt rejected the idea that the wounded and the family of the miners who were killed should receive any compensation for their loss (Hoyt 1898).

The Austro-Hungarian government tried several times to get the United States government to pay some restitution to the victims and their families, not based on the court case, but rather based on the principles of charity and humanity (Turner 1984, 145). The opinion of the United States solicitor, W. L. Penfield (1913, 267), was addressed to the secretary of state in January 1899. The solicitor discussed the claim and recommended that it not be granted.

The Growth of a Multiethnic Union

After the trial, Fahy began organizing in the Wyoming-Lackawanna field (northern anthracite field). Initially, the better-off "English speaking" miners were reluctant to join; however, the new immigrants joined without hesitation. Fahy named Slavs and Italians to top posts in the locals. A five-month successful strike against the Susquehanna Coal Company near Nanticoke (outside Wilkes-Barre) in the second half of 1898 proved the power of organizing the new immigrants. In mid-1899, Fahy returned to the Schuylkill fields, where he attempted to organize in 1894 and 1895. He named a Polish American as his assistant and the organization grew rapidly (Miller and Sharpless 1998, 245; Greene 1968, 158).

In 1899, under the leadership of John Mitchell, president of the UMWA (from 1898 to 1908), the organization made a strong recruitment effort in the anthracite region. By 1900, the UMWA began developing comprehensive strategies, rather than reacting to local disputes. Mitchell developed a cohesive organizational framework, which was disciplined and created a solid front (Yellen 1936, 143–44). "As a result, the UMWA succeeded in uniting fragmented local organizations and limiting the interregional discord and violence that had plagued the region's labor movement in the past" (O'Bannon et al. 1997, 62).

Mitchell did not feel the union was strong enough to negotiate for a wage increase, nor did they have the resources for a sustained strike. However, in 1900 the bituminous workers in western Pennsylvania received a 20 percent wage increase. The executive board sent Mitchell to seek an agreement with the anthracite coal operators (Miller and Sharpless 1998, 250). The coal operators declined to meet with Mitchell and they refused to recognize the union. On September 17, 1900, the UMWA issued a call to strike. John Mitchell spoke out for a more inclusive UMWA when he exclaimed at the beginning of the 1900 coal strike that "The coal you dig isn't Slavish or Polish, or Irish coal. It's just coal" (quoted in Beik 2002, 67). This phrase became the rallying slogan for the 1900 strike, as well as for subsequent actions.

About 65 percent of the mineworkers were UMWA members, and 125,000 men went on strike. Mary Harris "Mother" Jones joined the protest. Harris, along with other strike leaders, went from colliery to colliery closing operations. In October 1900, Republican politicians told coal operators that they needed to meet some of the union's demands and end the strike immediately, if they wanted to avoid Republican losses, including President William McKinley's reelection bid in the November elections. Republicans also feared that the strike would spread to the bituminous coal fields in western Pennsylvania. J. P. Morgan, who had significant financial interests in anthracite, also urged the operators to negotiate with the miners. The strike ended on October 29, when the coal companies conceded to the union's wage demands. The workers also gained the eight-hour day. John Mitchell's stature grew in the national public view. He was seen as a skillful negotiator who was willing to compromise. The miners dedicated October 29 a holiday and named it "John Mitchell Day." However, the coal operators still did not recognize the union or John Mitchell (Cornell 1957, 39–40; Dulles and Dubofsky 1984, 180–81). The day is still commemorated in the region among labor activists and miners.

In March 1902, the UMWA met and voted to strike. On May 12, Mitchell called for a temporary suspension of work in the mines. However, three days later the miners met at Hazleton and voted to continue the strike. More than 147,000 miners walked off the job. It was the beginning of one of the greatest strikes in American history. Miners asked for the same terms as two years earlier, including wages that equaled those found in the bituminous region, better safety conditions, and freedoms from the company store (Foner 1964; Fink 2015, 55).

During the strike George Baer, the Philadelphia and Reading Railroad's president, wrote a letter to a Wilkes-Barre clergyman who was sympathetic to the strikers. The letter became public and made history. Baer wrote:

> I see that you are a religious man; but you are evidently biased in favor of the right of the workingman to control a business in which he has no other interest than to secure fair wages for the work he does. I beg of you not to be discouraged. The rights and interests of the laboring man will be protected and cared for—not by the labor agitators, but by the Christian men of property to whom God has given control of property rights of country, and upon the successful management of which so much depends. (quoted from Reynolds 1960, 95)

Baer's letter was released to the newspapers. His Divine Right argument created a backlash and swayed public opinion away from the coal barons and toward the strikers. Newspapers, like the *Chicago Tribune* and the *New York Times,* weighed in against the operators. In September 1902, the New York State Democratic Convention met and called for government ownership of the anthracite mines (Miller and Sharpless 1998, 276).

President Roosevelt intervened and encouraged arbitration. The operators agreed to arbitration by a presidential commission, known as the Anthracite Coal Commission, and the miners went back to work on October 23, 1902, after striking for 163 days. According to Mitchell and members of the UMWA, capital and the government officially recognized the labor union. The *United Mine Workers Journal* reported on October 30, 1902 that a convention of seven hundred delegates voted in favor of arbitration by the commission. At the end of the convention, they all sang "My Country Tis of Thee." The journal reported, "A score of races mingled in the hall—Saxon and Celt, Teton and Slav, Latin and the native-born—and all joined. One-third of them could not follow the words or melody, but they knew that the hymn was of America, and their hearts sang. The echo of the chorus may well ring through the land, for it is a greeting of free men to a larger freedom that they won for all" (*United Mine Workers Journal* 2007, 5).

The 1902 strike was significant for three different reasons. First, organized workers won a historic victory over one of the most powerful anti-union capitalists. Second, during a crucial moment of the strike, the workers had the U.S. government siding with the union. Third, it was the first time that a massive strike did not have the American public condemn the actions of labor as a red menace (Fink 2015, 54).

Decline of Anthracite

The benefits of an uneasy peace between labor and capital since the 1902 coal strike became evident, as there was a steady rise of coal extraction. In 1901, miners removed about sixty-seven million tons of coal. This output grew steadily over the next two decades, and by 1917 anthracite coal extraction rose to about one hundred million tons and employed 180,000 workers. During World War I, the coal industry had a difficult time meeting demand. However, after the war, oil and natural gas gained an increasingly larger share of the market and the anthracite industry began its rapid decline (Rose 1981, 77). By 1922 the production of coal had dropped by nearly 40 percent. Rather than thinking about innovation and other uses for anthracite, the corporations wanted to maintain their wartime profits and cut costs by decreasing wages. These actions led to several work stoppages in 1922, 1923, and 1925.

By 1938, the industry employment rolls decreased to about 97,000 workers, and by 1950 there were about 50,000 employed in the industry. Throughout the decade of the 1950s, unemployment in the region never fell below 10 percent (Rose 1981, 104). A clear outmigration occurred at an increasing rate, and many towns' populations decreased by half or more. The 1959 disaster at the Knox Coal Company continued to damage the reputation of the anthracite coal industry. In this case the miners were instructed to tunnel upward, within six feet of the Susquehanna River, when the minimum safety regulation required thirty-five feet. The flooding Susquehanna River broke through rock strata, damaged about two dozen mines, and killed twelve workers in what is now known as the Knox Mine disaster. As a result, underground mining ceased in the Wyoming Valley. A subsequent court case related to this disaster proved that the company was not following safety regulations. Ten people were indicted, including the mine superintendent, the owner, and the secret owner, who was the president of District 1 of the UMWA, and three union officials. Six served time in jail for their illegal handling of the mine operations. By 1980, about three thousand people were employed in the industry, and twenty years later only about one thousand workers were employed in anthracite mining (Miller and Sharpless 1985, 321).

The current commercial market for anthracite coal is small, and a few companies are now selling anthracite in forty-pound bags for domestic use. It ends up being cheaper than oil or natural gas to heat a house. However, because of increasing environmental concerns and ordinances in major cities prohibiting the burning of coal for home heating, the market has decreased significantly. In 1998, the mines in the anthracite region produced more than five million tons of coal, and about a decade later the production fell to about 1.6 million tons. The output continues to decrease (*United Mine Workers Journal* 2007, 5).

The strip mines around Lattimer that were active in the early 2010s are now quiet. Some are in the process of being reclaimed, while others are waiting for a rebirth promised by politicians in the face of growing competition from natural gas production. To many the abandoned strip mines invoke a memory of the once-powerful coal industry. To others it is a reminder of the contested relationship between labor and capital. The confrontations over the control of the meaning and memory of coal's heritage occurs on many different fronts. The struggle for meaning and memory of the Lattimer massacre occurred almost immediately after the event. From the end of the trial, labor worked to memorialize the Lattimer massacre as a reminder of the unbalanced relationship between capital and labor. The contested memory between labor and capital was fought on the anthracite landscape over the next century.

CHAPTER 4

Memory of Lattimer

Remembering the Lattimer Massacre

Lattimer is the largest labor massacre in nineteenth-century America, yet it has been mostly forgotten in the national public memory. What aspects of the past are remembered and how they are remembered are important issues that allow us to see how places are commemorated and historicized. In 1925, Maurice Halbwachs formally introduced the concept of collective memory. He wrote that memories are located in a spatial and social framework. The development of collective memory originates in a set of individual recollections by a number of people who have a shared experience. Halbwachs (1992) remarked that a "collective memory" develops when individuals seek the testimony of others to validate their interpretations of their own experiences that provide independent confirmation (or refutation) of their memories, and thus confidence in their accuracy. Memories can be individual or collective as they mediate between the past and the present.

Memories can be "experienced" in several ways, through actions or gestures, monuments, and artifacts (Alcock 2002; Benton 2010; Meskell 2003). Memory often presents an uncomplicated way of seeing the past through deeds and acts that are often simple, thus allowing many to remember what they want to. A material and historical perspective of an anthracite region's landscapes provides some fascinating stories related to labor, migration, and the struggle for representation in the official American memory. The control of a group's memory is usually a question of power. Individuals and groups frequently struggle over the meaning of memory as the power elite often imposes the official memory (Teski and Climo 1995, 2). Written more than a generation ago, Eric Hobsbawm's words still ring true about the context of memory in relationship to power. He writes, "The history which became part of the fund of knowledge or the ideology of nation, state or movement is not what has actually been preserved in popular memory, but what has been selected, written, pictured, popularized and institutionalized by those whose function it is to do so" (Hobsbawm 1983, 13).

Memory is important in creating national histories and inspiring nationalism (see, for instance, Howe 1999, 222–39; Nora 1999). Nation-states tend to be rooted in tradition, and this memory of the past appears to be "so 'natural' as to require no definition other than self-assertion" (Hobsbawm 1983, 14). Official histories of a nation require consensus building and the construction of a history from multiple, often conflicting memories (Anderson 1991, 163–85; Kohl 1998, 225; Scham 1998, 301–8; Trigger 1989).

Influential works by David Lowenthal (1985), Michael Frisch (1990), David Glassberg (1990, 1996), Michael Kammen (1991, 1997), John Bodnar (1992), Edward Linenthal (1993), and David Blight (2001) have guided public historians into addressing issues that show the connectedness between memory and power. Their work shows that we cannot assume all groups, and all members of the same group, understand the past in the same way. The same historical and material representation may have divergent meanings for different audiences (Glassberg 1996, 9–10; Lowenthal 1985), and there is often a struggle to create or subvert a past by various competing interest groups (see, for instance, Neustadt and May 1986; Peterson 1994).

Public memory is often about seizing political control of the narrative in which subaltern views are often suppressed or relegated to peripheral positions within the larger narrative. New and more encompassing narratives are often difficult to negotiate. Understanding the meaning of heritage and knowing the actors involved in the creation of the meaning of heritage is important for situating a heritage project in a community (Brunner 1986, 142–46; Martin and Wodak 2003, 6).

A memory becomes public when a group has the resources and power to promote a particular past. Through material culture such as memorials, museums, and the built landscape, these histories can mask or naturalize inequalities. The public memory associated with highly visible objects is always being constructed, changed, and challenged. While there is always a strong movement to remove subordinate memories from the national collective memory, minority groups continually struggle to have their histories remembered. The clash over the control of public history occurs in some of the most visible places on the landscape, like national monuments and parks. These are the arenas for negotiating meanings of the past (Boric 2010; Jones 2007).

Many of the studies in memory fall into two categories: forgetting about or excluding an alternative past, or creating and reinforcing the hegemonic structure in society. People experience, remember, or forget collectively, in addition to sharing interpretations of these events. Ritual is one way of reinforcing and controlling the memory of the past (Shackel 2001a, 2001b). While monuments and markers on the landscape are designed to convey a certain memory, the process of repetitive ritual acts to reinforce, control, or even change this memory.

Memories can be public as well as private, and they serve to legitimize the past and the present. Elements of the past remembered in common, as well as elements of the past forgotten in common, are essential for group cohesion. The past of some groups may be rejected in order to create ties with other pasts, often for the political and social advantages of creating new identities (Benton 2010).

The control of memory can reinforce a hegemonic past and present. The official expression sponsored by the state is concerned with promoting and preserving the ideals of cultural leaders and authorities, developing social unity, and maintaining the status quo. It is a way to help reduce competing interests. Cultural leaders orchestrate commemorative events to calm anxiety about change or political events, eliminate citizens' indifference toward official concerns, promote exemplary patterns of citizen behavior, and stress citizen's duties over rights. They do this because of the existence of social contradictions, alternative views, and indifference that perpetuate fears of societal dissolution and unregulated political behavior (Van Dyke and Alcock 2003).

Michel-Rolph Trouillot (1995, 15) provides a compelling argument about the memory of historic events in Haiti. He notes that selective remembering often means that some accounts are left out of the dominant narrative. The history that is not silenced is what the general narrative of the past becomes. Not all sources are valued the same throughout a community, and to some extent the dominant narrative often has components of a fictitious story. Trouillot cautions us to be aware of the silenced past.

William Logan (2012) notes that memory and cultural heritage should also consider cultural diversity and enforcing human rights. He cautions us that official heritage interventions, which may be performed to achieve political goals, can also undermine rather than strengthen community identity, cultural diversity, and human rights. Though it is imperative that we critically think about the way heritage is created and the political purposes it serves, I also believe it is important to think about how we can mobilize our heritage work and address social justice issues (Shackel 2013; Little and Shackel 2014).

The story about the Lattimer massacre is about the struggle to remember and commemorate the event. A type of amnesia or forgetting is associated with the event, and this forgetting is a political act. There are varying and competing narratives that struggle to control the discourse of the event. It was in the interest of the coal operators to not promote or remember the incident, hoping to reinforce a docile workforce. Many of the miners who were known to be associated with the event were no longer welcome to work in the mines. About two weeks after the incident, the *Daily Standard* (September 27, 1897) reported that a "Hungarian" who appeared in the preliminary hearings as a witness was fired, along with about fifteen other employees. After some of the miners returned from the trial in March 1898, they were blacklisted from employment in the mines. "When I returned home," one witness related, "I went to John Beach, outside foreman at the A. Pardee & Co.'s Harwood mine, where I have been employed for two years, and told him that I was ready to go back to work. 'You were on the stand against us,' said Beach, "and told a lot of lies. You are no good. You get out of here and go, you can never work at this colliery'" (quoted in Pinkowski 1950, 34). Retribution and blacklisting appeared to be common. Another witness who was discharged said that the mine foreman told him: "There will be a lot of jobs and a great many vacant company houses around here when this trial is over" (quoted in Pinkowski 1950, 34). In a 1977 oral history, the son of one of the striking miners who participated in the Lattimer march said that his father and his brothers escaped death "as they lay down in a ditch during the shooting. . . . The men were so afraid of the company at that time that they did not like to talk about it. In a way, they hid their involvement" (interview summary with John Tomasko, August 6, 1977). They were afraid of retribution. One blogger to our Web site (lattimermassacre.wordpress.com) explained that his great aunt, Grace (Coyle) Nugent, who was the local schoolteacher at the time of the massacre, was also blacklisted for aiding the wounded at the site. "Grace was also 'Blackballed' from teaching in PA, as a result of coming to the aid of the mortally wounded miner, Andrew Jurechek. Grace ripped her petticoats and tried to bandage Andrew's stomach, pretty much destroyed by the bullets, and his only desire was to see his wife before he died and then right in front of Grace" (Finan 2013).

Soon after the acquittal of the sheriff and his deputies, Martin assigned eight of his men to guard the colliery and patrol the streets of Lattimer with rifles in hand. The *New York Evening Journal* (March 16, 1898) noted that in all other coal towns, men guard the company's property; however, in Lattimer their assigned duty was to protect the company property and the lives of the company's witnesses. Sheriff Martin acknowledged their presence; however, he would not admit if they were deputized. The town or the township did not authorize their presence. There were also fights between those sympathizing with the miners and those who sympathized with the deputies. The animosity escalated to stabbings and even arson. Coal operators were enraged when James Roderick, superintendent of the A. S. Van Winkle & Co., donated $100 to support the widows and orphans of the Lattimer victims. They asked the coal operator to dismiss him. Van Winkle refused, although Roderick was rebuffed by the other superintendents and coal operators at future meetings and gatherings (*New York Evening Journal,* March 17, 1898).

Grief is another reason for the widespread amnesia related to the event. Referring to her mother, the daughter of Michael Cheslock, Anna explained; "She also told her children that their grandfather was shot between the eyes, as she recalled the sight of the bullet wound and the blood gushing down his face. The hole in his head kept draining" (interview summary with Anna S., January 17, 1973). Massacre victim Michael Cheslock's great-granddaughter recounts:

> "It was almost like it was something . . . too bad . . . too horrendous to talk about. Nobody relayed anything about the story. . . . The closest of the daughters [of Michael Cheslock] was Anna, our Aunt Anna who lived in Freeland [PA], and we were over there . . . every couple of Sundays . . . and even she never related anything about the Lattimer massacre . . . it never came up in conversation. We never asked. (Interview, March 19, 2010)

Slavs began to leave the region in increasing numbers after Lattimer, many returning penniless to their native country. Their hopes of making a better life were dashed. The *New York Evening Journal* of March 17, 1898, explains, "They will go back to drudgery, almost slavery, but under the monarchy government of Austria there will be no Lattimer, and there will be some justice." Jurors and witnesses who testified in the defense of the deputies, if working for coal companies, were often promoted. Those not employed found jobs in the coal companies (*New York Evening Journal,* March 17, 1898).

Lattimer did not receive the same amount of news coverage as other contemporary labor disputes. Following the trial and acquittal of Sheriff Martin, the outcry over the killing of the miners quickly fizzled. Less than a month after the trial the UMWA (*United Mine Workers Journal,* March 17, 1898, 4) wrote a response to the verdict:

> Since the indictment of Sheriff Martin and his deputies for the shooting down of the striking marching mine workers at Lattimer . . . the JOURNAL has refrained from mentioning the case, preferring to wait until trial was had and the verdict rendered. . . . We cannot see any reason to suppose that this verdict was based upon the facts and evidence in the case, but believe it was based and justified by a desire to uphold the police power at any cost. . . . However, the verdict has been rendered and no amount of opposition thereto can change it. . . .

Perry Blatz (2002, 50) notes that the *New York Times Index* has more than 150 articles written about the Homestead strike in the latter part of 1892, while Lattimer has only twenty articles in the latter part of 1897. Lattimer, however, did receive significant coverage in some of the foreign-language newspapers printed in the United States. The Homestead strike is associated with two of the biggest capitalists of the era, Henry Clay Frick and Andrew Carnegie, and it occurred in booming Pittsburgh. The coalmine operators around Hazelton were less known, and the region is outside the influence of a major urban area (Blatz 2002, 50).

Memorialization

Memorialization efforts began soon after the end of the trial. In April 1898, the *Daily Standard* began reporting on efforts to create a monument. It noted that it could not be placed in Lattimer because the land was still owned by the coal company. However, the article did mention the possibility of placing it in St. Stanislaus cemetery, where fourteen of the miners are buried (*Daily Standard,* April 22, 1898). At the one-year anniversary of the massacre, a memorial service occurred at St. Stanislaus church, as well as a demonstration, where about 1,500 to 2,000 miners paraded through town in remembrance of the martyrs. Speakers at the event did not focus on the clash between the sheriff and the workers, but rather they spoke about the benefits of organized labor (*Daily Standard,* September 6, 1898, and September 9, 1898; *Philadelphia Inquirer,* September 11, 1898; *Plain Speaker,* September 12, 1898).

The Lithuanian-Polish Club proposed to raise $5,000, and the organization donated $500 to the effort. A design was proposed, an obelisk that would be about fifteen feet high, with the names of the victims inscribed on it as well as the shooting date (*Plain Speaker,* March 10, 1902). While those associated with the striking miners of Lattimer found it difficult to find employment in the region, many tried to disassociate themselves from the event. At the time, people generally considered employment opportunities more important than remembering a failed strike. Fund-raising for a Lattimer memorial continued. Though most of the anthracite region was providing donations for the monument, very few funds were coming from the Hazleton area (*Plain Speaker,* May 23, 1902).

After the successful 1902 Anthracite Coal Strike, the UMWA recognized that Lattimer was the turning point for the growth of the organization. In 1903, the UMWA locals collected over $5,000 to erect a monument to the miners shot at Lattimer. The design of the monument would be a thirty-foot-high granite shaft resting on a massive, ten-foot base of granite. There would also be a life-size bronze figure of a miner. The stone would have inscriptions and possibly the names of the dead miners (*Wilkes-Barre Semi-Weekly Record,* July 31, 1903, 7).

Wilkes-Barre was the chosen site because it is the county seat and considered centrally located. Those involved believed that the county seat would offer larger viewing opportunities than the smaller city of Hazleton or the coal-patch town of Lattimer. In addition, the coal company still owned the land where the massacre occurred, and they would not allow for the erection of the memorial (*Wilkes-Barre Semi-Weekly Record,* July 31, 1903, 7). There was some mention that the monument could be erected on the public square or on the river common. The *Wilkes-Barre Semi-Weekly Record* (July 31, 1903, 7) reported, however, that there was some opposition to the erection of the monument in the city, "as it will recall the deplorable labor troubles which it would be better to forget than to perpetuate in stone." If it was erected in town, the newspaper editor noted, it would spur a large demonstration of mineworkers from all over the county.

Mother Jones was to arrive at Hazleton to celebrate Labor Day and to commemorate the martyrs of Lattimer on September 7, 1903 (*Philadelphia Inquirer,* August 8, 1903). The *Philadelphia Inquirer* (September 8, 1903) mentioned that there was a big demonstration at Lattimer, followed by speeches by local UMWA union leaders. Mother Jones appears to have been absent.

The *Wilkes-Barre Times* (September 16, 1903) reported that enough money had been raised for a memorial for the Lattimer martyrs, and that one would be erected, without fail, in either Hazleton or Lattimer. The matter was now in the hands of John Mitchell, president of the UMWA. In 1907, the district presidents and vice presidents of the UMWA held a meeting to decide on the design and placement of the Lattimer memorial. Two plots of land had been purchased in West Hazleton (at the corner of Second and Broad Streets) by the UMWA for the placement of the monument, and the organization raised $10,000, more than necessary for the construction of an obelisk with a large base, to be ready for the tenth anniversary (*Wilkes-Barre Times,* August 24, 1907, and October 15, 1907). In April 1908, a design for the memorial was selected, at a cost of $5,000. It would consist of a thirty-foot granite shaft with a nine-foot base. At the top of the shaft would appear a miner in work clothes (*Wilkes-Barre Times,* April 17, 1908). The 1908 commemoration of the Lattimer Massacre occurred without the construction of the memorial (*Wilkes-Barre Times,* September 11, 1908). In 1911 and 1912, local newspapers published articles about the plans for the memorial (*Wilkes-Barre Leader,* September 22, 1911, and September 18, 1912). The city council was reluctant to support the placement of the memorial, especially on city property. If they supported the memorial, they would be condoning the actions of the "rioters," and if they rejected it, some would view them as supporting the shooting. More than a decade after the massacre, the *Wilkes-Barre Times Leader* (September 18, 1912) portrayed the event as a local matter rather than nationally relevant. The newspaper article noted that there were no monuments located on city open spaces that commemorate national struggles. Therefore, the paper believed that a local event should not be memorialized in the city. A better option would place it closer to Lattimer.

The monument was never erected, and the newspapers were silent about how the movement to erect a monument to the Lattimer martyrs fizzled. There is no mention of what happened to the money raised for the memorial. Not much is noted in the local newspapers about Lattimer and the memorial over the next several decades. In 1937, the *Sunday Independent* (June 20: Section 3, 1), a newspaper in Wilkes Barre, described that the event as the "Lattimer riot," and then took a pro-labor stance and reprinted part of an article published in the *American Weekly* and *Review of Reviews* from October 1897. The article states that the miners were unarmed, and the first volley of shots killed and wounded several of the strikers. The demonstration was broken up on the first onslaught of bullets. "Nevertheless the deputies kept on shooting at the helpless, unarmed, retreating men. . . . Indeed, the worst actions that they brought against the miners would, in a New York City strike, scarcely have justified the mild use of a policeman's billy." (*Sunday Independent,* June 20, 1937, Section 3, 1). The article later described that the miners were striking because of the sporadic and poor working conditions. There were no compensation laws, and the coal operators would rather smash the strike rather than try to address their grievances. At a legislative hearing at about the time of Lattimer, "pitiful tales were told of undesirable conditions, company managed and company owned stores, incredibly low wages, of home conditions that can hardly be visualized at this time" (*Sunday Independent,* June 20, 1937, Section 3, 1).

Commemoration and the Building of the 1972 Monument

In 1971, discussions developed once again about creating a memorial and a historical roadside marker to honor those killed at Lattimer. During the summer, the community held several meetings to discuss the plans for a seventy-fifth anniversary commemorative event to be held in 1972. The United Labor Council of Lower Luzerne and Carbon Counties, AFL-CIO supported the committee and provided them with some operating funds.

Albert and Amelia Cherko sold for $1.00 a parcel of land for the memorial to the United Labor Council of Lower Luzerne and Carbon Counties, AFL-CIO. The design for the memorial was approved, and a massive memorial rock came from the strippings (that is, open pit mine) northeast of Lattimer. Pagnotti Enterprises transported the memorial rock to the commemorative site, and they also moved a coal car to the site, which was donated by the Greenwood Mining Company (Lattimer Labor Memorial Year 1972).

The memorial was dedicated on September 10, 1972. It became known to some as "Labor's Rock of Solidarity," while others refer to it as the "Remembrance Rock" (Pinkowski 1997). According to the chair of the memorial committee, Charles McGlynn, it represents organized labor's eternal struggle to work, live, and die with dignity (Pavloski, September 9, 1993). On August 10, 1972, Governor Milton Shapp declared the year as "Lattimer Labor Memorial Year," and he called upon Pennsylvania residents to remember and appreciate the efforts of the coal miners who died at Lattimer. They "gave their lives unnecessarily in an effort to provide better living conditions for the coal miners of Pennsylvania" (Shapp quoted in Curilla 1972) (figure 4.1).

The monument's dedication was co-sponsored by the United Labor Council of Lower Luzerne and Carbon Counties, AFL-CIO, and the United Mine Workers of America. The United Labor Council of Lower Luzerne and Carbon Counties, located in Hazleton, also dedicated 1972 as "Lattimer

Figure 4.1. Lattimer Massacre Memorial, 1978. The Rock of Remembrance is to the right and the highway commemorative plaque is to the left. The "massacre tree" is still standing and is in the background. (Courtesy *Standard Speaker,* Hazleton, Pa.)

Labor Memorial Year." The AFL-CIO views Lattimer as a place to remember not only the martyrs associated with the event, but also to "remember the men and women of our trade union past" (*Pennsylvania AFL-CIO News,* February 1972). The ceremonies began with the unveiling of the stone, which had a large bronze plaque with the names of the Lattimer massacre victims, and a pick and shovel made of bronze at the base of the monument. The United Labor Council of Lower Luzerne and Carbon Counties, AFL-CIO and the UMWA sponsored the plaque. It reads:

> Lattimer Massacre, September 10, 1897.
>
> It was not a battle because they were not aggressive, nor were they on the defensive, because they had no weapons of any kind and were simply shot down like so many worthless objects, each of the licensed life takers trying to outdo the others in the butchery.
>
> Dedicated to these Union Brothers who made the Supreme Sacrifice.
>
> Sponsored by the United Labor Council of Lower Luzerne and Carbon Counties, AFL-CIO and United Mine Workers of America, the 10th day of September, 1972.

These sentences come from an article printed in the *Daily Standard* (September 11, 1897) seventy-five years earlier.

Adjacent to the Remembrance Rock and the bronze plaque was a state-sponsored roadside table developed by the Pennsylvania Historical and Museum Commission. It read:

> LATTIMER MASSACRE
>
> Seeking collective bargaining and civil liberty, immigrant miners on strike were marching in protest from Harwood to Lattimer. Here, on Sept. 10, 1897, they were met by armed deputy sheriffs. The ensuing affray resulted in the death of more than twenty marchers.

Representatives from various unions attended the event, including union members from throughout the anthracite region, a large group from the United Farm Workers Union of Philadelphia, and members from the International Ladies Garment Workers of Lancaster (*Pennsylvania AFL-CIO News,* September 1972).

In a message to the United Labor Council from George Meany, president of the AFL-CIO, he claimed it was important and necessary for the trade unionists to pay tribute to the martyrs at Lattimer. Their "courage and sacrifice became the foundation on which the American labor movement was built. The victims were new to the labor movement and new to America. They were of different backgrounds and different origins. But they stood unified in their belief in freedom and justice for all men" (Meany quoted in an unidentified news clipping, September 30, 1972).

Cesar Chavez of the United Farm Workers of America delivered one of the first speeches at the event. At the time of Chavez's speech, the Farm Workers and the Teamsters were in the middle of a jurisdictional dispute. At the dedication, flags and banners from both unions stood along the approach to the monument (Aurand 2002, 9). The tone of his speech focused on strengthening the labor movement—as a whole. From Chavez's speech notes (Cesar Chavez at Lattimer, 1972), we can see that he makes the connection between Lattimer and the current struggles of immigrant labor. He explained:

- Groups of workers in American today
- lives so closely parallel, lives of those miners
- they too are immigrants
- they too—have strange sounding names
- they too—speak a foreign language
- they too—trying to Build a Union
- they too—face hostile sheriffs and recalcitrant employers
- they too—had powerful employers
- they too—Are Non-Violent, As these men were

He said, "Let there be strength and unity in the ranks of labor throughout this land." Chavez continued, "Let there be only one voice . . . the voice of the working man and woman. Let there be only one Lattimer and for God's sake let there be no more Lattimers. Let there be peace, let there be justice and let there be love among all people" (quoted in *Pennsylvania AFL-CIO News,* September 1972). Then, referring to his organization, Chavez explained, "We know only too well of the hardship and sacrifice of these mine workers because together there is another group of workers who have things in common. They are the United Farm Workers, many immigrants, who want to make a decent living in the United States."

Robert McIntyre, treasurer of the state AFL-CIO, mentioned John Mitchell's connection to Lattimer. There is a John Mitchell statue in Scranton, Pennsylvania, and at the top of the memorial there are two hands clasping, and a note—"An eight hour work day." He spoke about a constructive labor movement and then connected Lattimer to Cesar Chavez's work. "That is what the United Farm Workers are also doing today. Seventy-five years ago, they were fighting for the same thing" (quoted in *Pennsylvania AFL-CIO News,* September 1972). Leonard Pnakovich, international vice president of the UMWA, also paid tribute to Chavez and noted that the struggles of the United Farm Workers are the modern counterpart of the UMWA. The low wages that were characteristic of the coalfields are what the farmworkers face. He noted, "Farm workers, too, face the guns and clubs of hired strikebreakers and police in the employ of anti-union employers" (quoted in *Pennsylvania AFL-CIO News,* September 1972). Pnakovich sympathized with their battle for economic justice. He spoke about the American dream the strikers had. They dreamed to better themselves, no matter their ethnic or religious affiliation. The strikers had a dream to better their standard of living and a dream that their children could attain a better standard of living and not be barred from upward mobility because of class privilege. Pnakovich described how oppressive political and economic systems at the turn of the twentieth century made a mockery of the American dream. The new immigrant "saw a wage level which, at best, permitted only a hand-to mouth existence. They saw working conditions that were oppressive and extremely dangerous. They saw a system of company housing and company stores that robbed the miner of the little that he was able to bring home for his long and dangerous day's work underground" (Pnakovich quoted in Curilla 1972).

Pnakovich explained that, because of language barriers and nationality, many of the new immigrants were discriminated against. All they asked for was a right to live and work in America under the protection of the Constitution, which guaranteed to all men that those rights were inalienable (Curilla 1972). Pnakovich explained that "To attain those rights, the victims of Lattimer and their fellow miners exercised the most fundamental of human freedoms, the right to be heard. They marched down the roads of this region. They marched to this spot and here, like so many of their brothers in the history of our trade union movement, they died under the bullets of those who would continue the oppression of the union spirit, and who, in the interest of profit, would destroy the most sacred possession of all men—life itself" (quoted in *Pennsylvania AFL-CIO News,* September 1972).

A eulogy by John C. Brennan, secretary-treasurer of Local 1361, Retail Clerks International Association, AFL-CIO, also detailed the importance of making the connection between the memory of Lattimer and the labor movement. Brennan claimed that "Lattimer is a Trade Union memorial" and therefore, he explained that it is easy to link the issues surrounding the Lattimer massacre with issues workers face today. Brennan spoke about the need for a humane workplace and for workers to be treated as human beings. He also explained that job satisfaction is a "God-given right." Therefore, the worker should share in the governance of the workplace and participate in any decision making that would affect the workplace. "I suggest to you that a work place which does not provide for spiritual/psychological satisfaction along with material rewards is not a work place which is desired from the standpoint of the worker, the industry, or the nation" (Brennan

1972). Congressman Daniel J. Flood attended and gave one of his signature flamboyant speeches.

At the time of the commemorative event, union leaders worked to make Lattimer a "labor" event, while organized ethnic groups were overlooked and not part of the planning committee (Aurand 2002, 9). Brennan, however, later wrote, "The historical relevance of Chavez and the Immigrant Martyrs of Lattimer is profound: Ethnicity, socio-economics, labor history, and certainly, not in the least Christianity" (Letter, Brennan to Novak, April 9, 1976).

Memorialization in the 1990s

The Reverend Joseph Ferrara, Pastor, St. Mary's Church, Lattimer, began the Liturgy of the Mass in 1973—at the Rock. Throughout the 1970s and into the new century, a Catholic Mass was held at the memorial during the anniversary of the massacre (figure 4.2). Clergy usually came from Lattimer United Methodist Church, St. Mary's Church of Lattimer Mines, St. Joseph's R.C. Church in Hazelton, and Holy Infancy Church of Bethlehem (Brennan 1978; *Citizen's Voice,* September 10, 1984; Pavloski, September 9, 1993; Aurand 2002). In an article for the *Standard Speaker* (September 2, 1992), Charles McGlynn wrote about the ceremonies held at the Lattimer massacre

Figure 4.2. Religious service at the Lattimer massacre memorial, 1993. (Courtesy *Standard Speaker,* Hazleton, Pa.)

memorial on its ninety-fifth anniversary in 1992. He noted the respect, hope, and solidarity he experienced at the commemorative ceremony. The event was widely covered in the local newspaper, the *Standard Speaker,* and the local cable company aired the service held at the site.

At the beginning of the 1992 service, James Hiza and Nicholas Sedon, United Labor Council officers, lighted twenty-six candles—nineteen for the men who died at Lattimer, six for those who died soon after as a result of their wounds, and one for all of the union members who have carried on the fight for the unions—"Labor's Unknown Soldiers." The choir of St. Stanislaus R.C. Church, Hazleton, provided the music accompanying the service, a task that they performed for many years. The Most Reverend Francis Di Lorenzo, auxiliary bishop of Scranton, celebrated the Mass. In his homily, Di Lorenzo compared the social and economic situation of ninety-five years ago to the present. He stated, "What is apparently clear to all of us is that these men were actors in a drama which is an eternal drama—an economic drama which we are going through again. Economics was a precipitating part of this violence." He added that "people do not trust their leaders to solve their economic problems. Many of our people are on the lower end of the wage structure. There is hidden poverty in our area and the distribution of wealth is still a problem." The bishop concluded his homily by stating, "The Catholic Church can do two things to help people on the local level, direct services, and advocacy—state and federal elected officials must be made to hear and recognize the needs of the poverty stricken" (quoted in McGlynn, September 2, 1992). During the offertory, where bread and wine is offered to the people gathered at the ceremony, symbols of labor were carried to the altar by retired mine workers and members of the United Labor Council. The gifts included a piece of coal, a miner's pick, a thermos bottle, and a plastic safety helmet (McGlynn, September 2, 1992).

After the services, UMWA president Richard Trumka addressed the audience from "The Rock." He asked the gathering, "Why have we held the line against our enemies? The reason is here in Lattimer. With all of the pain and hardships, one thing that stands out in all of our victories is our solidarity." Trumka explained, "Bigotry and distrust held the unions down. Then it was prejudice, today it is called racism. John Fahy, a union organizer of that era, told the miners, 'You can be Hungarian, Irish, Polish, Italian, Slovak, German, or Lithuanian, but you could also be union.' Thousands of miners joined the UMWA following the Lattimer massacre." Trumka added, "What our miners lack in riches and wealth we more than make up in courage. Working people standing together can win battles that cannot be won individually. Now, the only guarantee we have of victory is our own solidarity" (quoted in McGlynn, September 2, 1992).

McGlynn concluded his article by stating that the commemorative events were important in many ways. The ceremony was about remembering the martyrs of Lattimer and the importance of solidarity, the mortar that holds unions together. It was a way to show that organized labor has compassion and remains alive. Also, top officials came to the event reflecting on their respect, hope, confidence, and solidarity. Hope and confidence exists that the future will produce "responsible unionism and the glory days of this country" (McGlynn, September 2, 1992).

Making a broader social and political statement about the importance of remembering Lattimer, McGlynn explained that the lessons of the past, the ability of various ethnic groups to organize together, is an important lesson for us today. "The Lattimer massacre memorial is a symbol that tries to include all races, creeds, and ethnic groups. The memorial includes the idea that different ethnic groups are in solidarity, and the reason they are in solidarity is because they march under the banner of labor" (quoted in Pavloski, September 9, 1993).

The *Citizen's Voice* article identified all of the clergy who would be present at the memorial services on September 10, 1993. It is interesting that the author included the pedigree and connection of each of the clergy to the region's coal mining industry. "Rev. Bucolo's father was a coal miner who worked for 40 years at Jeddo 4 and Harleigh. Rev. Garbacik's father also was a coal miner about 40 years at Cranberry. Rev. Matz's father also was a coal miner. Rev Grabish is the son of a coal miner who grew up in nearby St. Clair. Rev. Angelo's father was a carpenter, while Rev. Grabish has attended the services for many years" (Pavloski September 9, 1993).

Father Garbacik, pastor of St. Stanislaus R.C. Church in Hazleton, realized that four of the graves in St. Stanislaus cemetery were unmarked, and he purchased four memorial stones, adjacent to the ten memorial stones that had been placed along the cemetery wall decades before. The four new additions included stones for Andrew Kulik, Andrew Szymanski, Anthony Gryczka, and John Tarnowicz. They were not included among the original grave markers because their death certificates are located in other churches, although they were buried at St. Stanislaus cemetery after the first ten men were buried in a mass grave. Pinkowski claims that Father Garbacik personally mowed the lawn around Father Aust's grave site many times, as a way to remember his generosity for helping the families of the miners and helping to raise money for the trial in Wilkes-Barre (Pinkowski 1997) (figure 4.3).

At the 1993 commemorative service, the Rev. Joseph Bucolo, pastor of St. Mary's R. C. Church, Lattimer Mines, in his homily spoke about the deplorable conditions miners and their families faced. He explained that they were striking for basic rights to work and live with dignity. The memorial at

Figure 4.3. Burial site of fourteen Lattimer martyrs in St. Stanislaus Cemetery, Hazleton. (Photograph by Paul A. Shackel, 2012.)

the entrance to Lattimer, Bucolo explained, is a symbol of organized labor's continuing struggle to work, live, and die with dignity (Moisey, September 11, 1993).

Ethnicity vs. Labor

The struggle to control and define the meaning of Lattimer continues. From at least the 1970s Edward Pinkowski, author of the first monograph (1950) of the Lattimer massacre, claimed that the state-sponsored roadside marker developed by the Pennsylvania Historical and Museum Commission in 1972 recognized the heroic work of the striking miners; however, it ignored the ethnic origins of the miners. In 1973, he wrote, "I want the Pennsylvania Historical and Museum Commission to take down the historical marker it erected at Lattimer because it maligns the 19 men who died there for the cause of unionism" (Letter, Pinkowski to Novak, January 12, 1973). While Pinkowski argued for the representation of all of the miners' ethnic backgrounds, Carlo DeMarco, former United Labor Council president, insisted that the marker and the yearly commemorative celebrations at the memorial were designed to recognize the labor movement without regard to ethnic background of the massacre's victims (Gloman 1989, 9). DeMarco stated that the marker memorializes the tragic fate of workers who were protesting for 1) better working conditions; 2) the abolishment of unfair company stores; and 3) higher wages for their work. DeMarco insisted, "We are a labor group and we were memorializing a labor movement. In no way did we intend to ethnicize that event, nor did we extend any preference to nationality except to mention that the victims were of Eastern European or Slavonian descent" (DeMarco quoted in Gloman 1989, 9). DeMarco also accused Pinkowski's efforts as "attempting to defame the Lattimer massacre event" (DeMarco quoted in Gloman 1989, 9). DeMarco has been placing American flags at the graves of the martyrs at the Stanislaus Cemetery site. "In doing so I'm honoring men who gave their lives for the same cause as any American. Their nationality is of no significance. Their purpose and goal are what count" (DeMarco quoted in Gloman 1989, 9). Meanwhile Msgr. Francis J. Breea, pastor of St. Joseph's R.C. Church, the oldest Slovak Catholic church in the western hemisphere, petitioned the PHMC to change the roadside markers at Lattimer to include the ethnic backgrounds of the martyrs. He and Pinkowski petitioned the PHMC to add satellite markers at the four cemeteries where the victims are buried. DeMarco, exasperated, concluded, "We would hope, therefore, that Mr. Pinkowski would not undermine that noble effort made by men who sought democracy, by littering up the true purpose of the Lattimer Massacre memorial with ethnic slurs" (DeMarco quoted in Gloman 1989, 9).

Pinkowski explains that the nineteen men who died at Lattimer were martyrs to freedom. However, he criticized the labor committee for not making a statement about what made the deputies fire upon the marchers. In an editorial in the *Standard Speaker* (1993), he claims, "It is absurd to pretend that overzealous deputies were not on a rampage to do a little ethnic cleansing under the guise of law and order" (Pinkowski 1993). He also stated that if the mine operators did not own the patch towns, the deputies might have burned down their homes. However, Pinkowski claims that the memorial committee has much unfinished work. "Never, to my knowledge, has the labor committee laid wreaths and paid homage to the Lattimer martyrs on their graves, in stark contrast to the laborites in Allegheny County who visit the graves of Fannie Sellins and Joseph Strzelcki instead of the hellhole where they were brutally shot" (Pinkowski 1993). Pinkowski ends his editorial stating that it is important to recognize the names and ethnicity of those who fell at Lattimer. It would serve as a reminder of the past and prompt people to think about living in freedom without bigotry, and without violation of human rights.

Charles E. McGlynn was, for a long time, the groundskeeper and the spiritual protector of the Lattimer Massacre Memorial. To him, the memorial is about one word—"dignity." McGlynn (quoted in Franklin 1990, 41) explained, "It's a dangerous word for many vested interests in this country. Even today, there are companies who see to it that ordinary workers don't find it in their work. This memorial is about that. It's a symbol for the workers not to forget about their dignity." He explained that the struggle for dignity is an eternal struggle and every union contract is another skirmish in the larger war for dignity. McGlynn was recognized by the Pennsylvania Labor Heritage Council and received its Mother Jones Award in 1990 for his efforts to keep the memory of Lattimer alive (Franklin 1990, 41).

McGlynn sees the area around the monument as "sacred ground." He explains that the grounds around Lattimer are a symbol of "labor's unknown soldiers." There are many workers who died in obscurity and their memories should not be forgotten. "No one wants to recognize that these people ever lived. They lived on behalf of responsible unionism, even sacrificed down the years for it. Many times they stood up for what was right and many times, they paid for it. Many times, they had to choose between standing up for their principles, or giving in. And many who chose to stand up were later judged unsuccessful by society's standards" (McGlynn quoted in Franklin 1990, 41). McGlynn (1994) later commented that the Lattimer massacre is also internationally significant because the outcome spurred the growth of labor unions.

Organized labor, according to McGlynn, was successful in creating the forty-hour workweek, workers' compensation, health insurance, and social security. "Every benefit worthwhile can be traced to organized labor. Who else would have pushed for these?" (McGlynn quoted in Franklin 1990, 41). The yearly Mass held at the massacre site has significant meanings on many different levels. "First, it is dedicated to 'Labor's Unknown Soldiers.' These are the many unknown men and women who dedicated their lives on behalf of responsible unionism. They will always be unknown and unseen, except in the eyes of God. Second, the Mass is dedicated to all of the miners in the anthracite as well as bituminous coal region. Labor considers the grounds around Lattimer as of the most sacred grounds associated with labor's cause. The ground is saturated with the blood of the 'Martyrs of Lattimer' and the years of sacrifice and suffering of 'Labor's Unknown Soldiers' and all miners and workers" (McGlynn 1994).

In 1997, the *Standard Speaker* noted that Pinkowski called the original state historical marker at Lattimer "as obnoxious as a loud drunk in a crowded theater" (Funk 1997). Pinkowski believed that the generality of "more than 20" takes away from the individuals who died at the site, even though the statement "more than 20" includes the miners that died a few days later from gunshot wounds. Pinkowski stated, "Time has not brightened the prospects of restoring the ethnic roots of the 19 Lattimer martyrs. Will it take another century to recognize that Polish, Slovak and Lithuanian martyrs who met their deaths in 1897 because they could not speak English and did not understand Sheriff James Martin" (quoted in Funk 1997).

By the centennial commemoration at Lattimer, the event incorporated both the memory of labor history as well as ethnic history. A brochure (Wyoming Historical and Geological Society, 1997) outlining the events billed it this way: "The Lattimer massacre is one of the most significant events in American ethnic and labor history. [The commemorative events] will reaffirm the historical significance of this event and examine ways in which the memory of Lattimer continues to intrigue labor and ethnic groups today." Subsequent commemorative events began to broaden in scope and included the commemoration of the various ethnic groups as well as being inclusive of the various contemporary religions.

CHAPTER 5

The 1997 Centennial Commemoration and the Memory of Lattimer

The Centennial Event

The planning for the 1997 centennial commemoration of Lattimer included a committee that consisted of labor organizers; members represented different ethnic groups, as well as representatives from Pennsylvania Labor History Society, the Pennsylvania Historical and Museum Commission, the Eckley Miners' Village Museum Associates, and the Wyoming Historical and Geological Society (Aurand 2002). The different factions continued to debate about whether Lattimer was about labor or if it was about ethnicity.

On September 10, 1997, Congressman Paul E. Kanjorski from the district read a statement in the House of Representatives into the *Congressional Record* honoring the one-hundredth anniversary of the Lattimer massacre. In his statement, Kanjorski describes the events leading up to the massacre and explained that "These men forever changed the face of the American labor movement." In his speech he included the pro-labor inscriptions from the monument: "It was not a battle because they were not aggressive, nor were they on the defensive, because they had no weapons of any kind and were simply shot down like so many worthless objects; each of the licensed life takers trying to outdo the others in butchery." Kanjorski included an article that appeared in the *Standard-Speaker* to be part of the *Congressional Record* (Kanjorski 1997).

Pinkowski (1997) wrote that Father Garbacik was not included in the centennial celebration and he "took it on the chin." However, on September 10, 1997, he held a Mass in St. Stanislaus Church in remembrance of Father Aust and the victims of Lattimer. Father Garbacik died in 2009 and is buried next to Father Aust. Garbacik supported and spearheaded the effort to erect a memorial stone at St. Stanislaus Cemetery that was placed on the cemetery's boundary wall. The Lattimer martyrs' grave markers are

Figure 5.1. Roadside marker erected on Rt. 924 near Harwood, Pennsylvania. Since its placement in 1997, other roadside advertisements have been placed around the commemorative marker. The marker now blends in and become less effective with the other roadside signs. (Photograph by Kristin Sullivan, 2010.)

adjacent to the memorial stone and located in the interior of the cemetery wall.

The formal centennial events occurred on September 12 and 13.* On Friday, September 12, there was a dedication of a state historical marker near the place where the march began in Harwood (figure 5.1). The marker is located on Route 924, near the entrance to Harwood, and it refers to the nationalities of the marchers as well as the cemeteries in which they are buried. It reads:

---

* The sponsors for the centennial events on September 12 and 13 included: Area Polish Cultural Club of Mount Carmel; 1st Catholic Slovak Ladies Association; 1st Catholic Slovak Union; Greater Hazleton Area Polonaise Society; Greater Wilkes-Barre Area Labor Council; Knights of Lithuania; Ladies Pennsylvania Slovak Catholic Union; the State Department of Labor and Industry; the State Federation of Museums and Historical Organizations; the State Heritage Society; the State Historical Association; the Polish American Association of Harrisburg Inc.; the Slovak Catholic Sokols; the Union of Needle Trades, Industrial and Textile Employees (UNITE!); United Labor Council of Lower Luzerne and Carbon Counties; and the United Mine Workers of America (UMWA) (Campomizzi-Clews 1997).

> Near here at Harwood, on Sept. 10, 1897, immigrant coal miners on strike began a march for higher wages and equal rights. Unarmed, they were fired upon at Lattimer by sheriff's deputies. Nineteen marchers—Polish, Slovak, and Lithuanian—were killed. The majority of the dead were buried in St. Stanislaus Cemetery, Hazleton. Others were interred in St. Joseph's & Vine Street Cemeteries. Hazleton, and in St. Patrick's Cemetery, McAdoo.

Brent Glass, the executive director of the Pennsylvania Historical and Museum Commission, unveiled the Harwood sign. People came to the unveiling ceremony for different reasons. State Senator Raphael "Ray" Musto explained that his father was one of the UMWA organizers for John E. Lewis. Musto was a Pennsylvania state senator from 1982 to 2010 and authored the state's Black Lung Act (quoted in Conrad 1997b, 1; interview with Brent Glass, May 10, 2010). Because of this legislation, Congressman Dan Flood was able to get a similar bill enacted at the federal level.

Representative Todd Eachus explained that the marchers were fighting for economic opportunity. What they were fighting for remains relevant today. "We've got to keep on fighting the same issues when there is a violation of basic justice" (quoted in Conrad 1997b, 1). Resident of Freeland, Pennsylvania, and former miner Philip Voystock explained, "I'm here today to honor my father, Frank, and his father Frantesek, who had come from Slovakia. Both my dad and grandfather died from miner's asthma. I came to show some respect for what they had stood for" (quoted in Conrad 1997b, 1).

In this region the Catholic Church is closely tied to the labor movement and social justice issues related to work and exploitation. Father Aust came to the aid of the wounded miners and the grieving families in 1897 and after. In recent times, Father Garbacik of St. Stanislaus Church (where fourteen of the victims are buried) and Father Ferrara of St. Mary's Church in Lattimer, among other clergy, supported the commemorative events surrounding Lattimer. Father Generose of Queen of Heaven Parish at Our Lady of Grace Church in Hazleton grew up in Lattimer and served in St. Mary's Church in Lattimer. He explains his reasoning and interpretation of the connection between labor and the church. "When you look at the social doctrine of Leo XIII from the late 19th century . . . and all these great encyclicals that they wrote during the industrial revolution, [they] are preserving the integrity of the human being from any kind of exploitation, and in the industrial revolution, we know what that was like for many people, ok, these people were [considered to be] subhuman. . . . Because God's love is for everyone, the church's understanding is that we're all called by God's gifts and the life that he has asked us to lead, to take the mission out of the church and into the world in which we live" (interview, Father Generose, August 20, 2014).

St. Mary's Church played a significant role in this centennial event. Cecil Roberts, president of the UMWA, led a commemorative march at 3:30 P.M. from St. Mary's Church through Lattimer to the massacre site (Beik 2002, 63). The marchers included a color guard and men dressed in mining clothes. With American flags waving, the group walked about a half mile from the church to the site of the massacre. Drums beat to the rhythm of the march and the group sang "solidarity forever, the union makes us strong" (Dempsey 1997, 3A). Also leading the march were several politicians (Congressman Kanjorski, State Senator Ray Musto, State Representative Todd Eachus), president of the Pennsylvania AFL-CIO William George, Cecil Roberts, and a reenactor dressed as Mother Jones. They arrived at the massacre site at 3:45, the same time the strikers arrived at Lattimer one hundred years earlier.

Once they reached the site, Roberts spoke of the miners' sacrifice: "These men did not die in vain. Somebody eventually would have to stand up as they did. Without the sacrifice of these workers that day, you and I would not be as well off as we are today. . . . They died for the noble right of man" (quoted in Conrad 1997c, 2). Congressman Paul Kanjorski said that words could not explain what the victims of Lattimer had contributed to the average working person. "We can never say enough, and, yet we still wouldn't be able to capture the essence of what had happened here 100 years ago," explained Kanjorski. "The issue that they fought for, and stood for, exists today all over the world" (quoted in Conrad 1997c, 2). Referring to the continued exploitation of workers, he told the crowd that there was no one in the crowd who was not wearing a piece of clothing made by an exploited worker (Conrad 1997c, 2). Then, representatives of each of the ethnic groups that participated in the Lattimer march placed a wreath at the foot of the memorial (Beik 2002, 63).

With continued pressure from Pinkowski and others, PHMC sponsored a new roadside marker at the site for the centennial event. It now reads:

> Here on September 10, 1897 nearly 400 immigrant miners on strike were met and fired upon by sheriff's deputies. Unarmed, they were marching from Harwood to Lattimer in support of higher wages and more equitable working conditions. Nineteen of the marchers were killed, and 38 were wounded. This was one of the most serious acts of violence in American labor history.

Despite the efforts of PHMC, Pinkowski (1997) noted that the state historical marker at Lattimer is incorrect. Not all of the men on the march to Lattimer were strikers. He describes the fate of a delivery boy, John Fotta, who was delivering beer to the coal-patch town of Harwood. He claims that before he could make his delivery, some of the striking miners halted his wagon, unloaded a keg of beer, and forced Fotta to drink with them. Intoxicated, Fotta joined the march to Lattimer, where Fotta lost his life. Even though Fotta was sympathetic to the miners' cause, Pinkowski brings up this detail to challenge the state-sponsored roadside marker's statement that all of the strikers were coal miners (Funk 1997). Though the new roadside marker was installed in time for the 1997 centennial, it noted that nineteen men were killed in Lattimer, which pleased Pinkowski; however, it does not refer to nationality.

On Saturday Michael Novak of the American Enterprise Institute delivered the keynote address at a round-table/panel discussion. Novak is author of the only book-length study of the event, *The Guns of Lattimer*. It is a historical novel with detailed historical research on the events that surround the Lattimer event. The historical fiction is easily separated from the historical research by chapters.

On Saturday evening, a Mass was held at the monument. The *Standard Speaker* (Tarone, September 14, 1997) wrote that two hundred people attended the service and it seemed to be a union rally. AFL-CIO and "Union Yes" flags stood behind the altar, and many people at the gathering sported union logos on their jackets. Bishop James Timlin conducted the full-service Mass, and in his homily he echoed Abraham Lincoln's Gettysburg Address. He proclaimed, "The world will remember what happened here 100 years ago, not what I say here today. Nineteen men were mowed down, killed, right here" (quoted in Tarone 1997). Timlin also made a connection to the place and the event by explaining his genealogy. He noted that his two grandfathers were miners in Scranton. His father was a union man, and Timlin felt great sympathy for the union movement. He explained that the miners who marched that day laid down their lives for the labor movement. Appealing to the union members, Timlin stated that "It was here where the UMW proved it was a force to be reckoned with" (quoted in Tarone 1997). He continued, "let us thank God for the men who came before us and the women who supported them. Remember, God can bring good things even out of bad things. What happened here 100 years ago was awful." He added that the good thing was that the plight of the American worker has improved markedly in succeeding decades, as a result. "Thank you for your solidarity," he told the gathering (quoted in Tarone 1997).

While the UMWA claims the memorial and the memory of the Lattimer massacre, many express mixed feelings about control over the memory of the massacre site. Some residents feel the UMWA has overstepped their boundary in claiming the site, because the majority of the workers striking were non-union miners and not UMWA members. "[The ceremony] got more unionized, you know, more union officials came in. . . . It wasn't like when we would go in the beginning" (interview with Gloria Cheslock and Mary Jo Cheslock Barrett, March 19, 2010).

Remembering Michael Cheslock

Many of the victims' families received some financial support for a limited time. Rev. Carl Houser, pastor of the Slovak Lutheran Church of Freeland, identified Cheslock's body and officiated the burial ceremony, which was held on the street in front of the Cheslock house in Harwood. Cheslock was then interred in the Hazleton Cemetery, also known today as the Vine Street Cemetery. Houser helped raise money for the family, and the Slovak Fraternal Society created and printed a remembrance card that shows the widow and children standing in front of their house and an inset of Michael Cheslock. The card sold for fifty cents, and the proceeds were donated to Mrs. Cheslock (Conrad 1997a, 34) (figure 5.2).

One interview with a Cheslock descendant explained,

> After his death, the widow and children lived a life of poverty. They existed on bread and coffee, and depended on handouts. (There was a collection taken up

Figure 5.2. A remembrance card that sold for fifty cents to raise money for Michael Cheslock's family. It shows his widow and children standing in front of their Harwood house, with an inset of Michael Cheslock. (Courtesy *Standard Speaker,* Hazleton, Pa.)

> for the survivors immediately after the massacre.) At the most, they had meat twice a year. The only assistance they received was $4 a month. They were looked down upon and mocked. At the time it was considered a big joke that there was a trial. Popular opinion was that there should have never been a trial. They were called "greenhorns." In addition to these difficulties, the language barrier was another constant barrier. Michael had been the only one in the family who really knew English. Mary told her children that people were always pulling things over on their grandmother because she could not understand the language. (Interview summary with Anna S., January 17, 1973)

Another interviewee noted, "There was not much concern about the widows and children because the men killed were from far away. The concern centered around the priest's prediction. . . . The priest from St. Stanislaus made a prediction that they all would die horrible deaths and everyone waited for it. . . . All the town waited to see whether or not he was right. None of the deputies died a natural death. None died without a struggle" (interview with John M., March 13, 1973).

Michael Cheslock is probably the most renowned figure of all of the martyrs associated with the Lattimer massacre. He was well respected for being an active member in the Slovak Lutheran Congregation in Freeland as well as other fraternal organizations. Cheslock was buried in the Vine Street cemetery in Hazleton, although his grave was never marked. In about 1986, Edward Pinkowski wrote a letter to the editor of the *Standard Speaker* noting that Cheslock's grave was unmarked and overgrown. A newspaper article appeared during the centennial celebrations, and Cheslock's granddaughter noted that he was buried in a "borrowed" grave near the "main gate" of the Vine Street Cemetery. However, because the family was poor, they lacked the resources to mark the gravesite. The article spurred John Probert, president of the Vine Street Cemetery Association, to begin looking for Cheslock's gravesite to memorialize him. In a search of the burial records for the cemetery, Probert found a ledger book that recorded receipts for digging graves. A notation read, "Sept. 11 Michael Ceslak 1 grave dig pd. $10.00." In another 1897 ledger book appeared the entry, "Sept 14 Mrs. Mike Cheslock, single burial place, digging grave for her husband $10.00" (quoted in Probert 1998).

Probert immediately contacted Pinkowski, who told Probert he had personally visited the gravesite with Michael Cheslock's daughter in the 1940s while researching for his monograph on Lattimer. Cheslock's daughter pointed out the gravesite to Pinkowski. It was inside the Alter St. gate and next to the burial site of Andro Rosman, whose grave is marked with a tall stone. Probert found a shallow depression to the right of Rosman's stone, a sure indication of a burial. Rosman and Cheslock were both Slovak

Lutherans and Cheslock was head of United Lutheran Society in Freeland. Apparently, the lodge arranged with the Vine Street Cemetery to reserve a number of single burial plots (SBP) in a row for its members. The cemetery records in 1895 show that Michael Cheslock paid $10.00 to bury Andro Rosman (Probert 1998).

Maurice Heater of the Heater Monument Co. donated the granite marker stone. The stone reads, "Lattimer Martyr / Michael Cheslock (Ceslak) / Born Oct. 17 1858 in Austria-Hungry / Immigrated to the USA Feb 13 1882 / Applied for Citizenship June 28, 1897 / Shot to Death by Sheriff's Deputies / In / Lattimer Massacre / September 10, 1897" (figure 5.3).

In an oral history, Dorothy German, the granddaughter of Michael Cheslock, described some of the incidents surrounding the massacre. They were told to her by her mother Mary, daughter of the Cheslocks, who was five years old when her father was killed. The Cheslocks had three other daughters—Ella, Verona, and Anna, and a son, John. John was the youngest, at ten months old at the time of the massacre. Mary retold the account of the day of the march from Harwood to Lattimer. "She said my grandfather was playing cards (pinochle) in a garden in Harwood when a group of men

Figure 5.3. Michael Cheslock grave marker erected in Hazleton Cemetery (Vine Street Cemetery) after the centennial of the massacre. (Photograph by Paul A. Shackel.)

showed up." Cheslock and his fellow card players were told that men from several collieries were marching to Lattimer in an attempt to get the men from the Lattimer mines to join their strike. "My grandfather told these men from McAdoo that he and his friends would join the march only if it was peaceful. They agreed only after they were assured none of the marchers would be carrying guns or clubs." Eager to join the march, "one man (Stephen Urich) had taken an American flag and he said he would lead the group. My mother said that when he moved to the front of the line with the flag, he shouted: 'We have nothing to worry about because we have the American flag to protect us'" (quoted in Conrad 1997a, 34).

German said she had witnessed an unforgettable scene. "I was in second grade (about seven years old) when a man came to our house one day and he asked to speak to my mother. He told her that he lived in Philadelphia but had been one of the sheriff's deputies who had fired on the marchers that day." She continued, "Then I remember seeing him get down on his knees in our kitchen and begging her for forgiveness. That's when my mother ordered me into the other room. After the man left my mother told us he told her he couldn't sleep because his head was filled with scenes of men lying in pools of blood—dead and wounded—and they just wouldn't leave him." Dorothy German continued, "She said he begged her to forgive him because, if she would do so, maybe he'd finally be able to get some sleep. My mother thinks he was the deputy who had shot my grandfather in the head." Dorothy German said that her mother did forgive the deputy for his role in the massacre. "But for years afterward, Mom would often tell us 'You can forgive but you can't forget'" (quoted in Conrad 1997a, 34).

Antagonist Views

By 1997, a commemorative stone slab erected by Father Garbacik was anchored on the wall of the St. Stanislaus Cemetery. It lay adjacent to and above the row of the fourteen martyrs' graves. The stone slab contains text facing the street implying the presence of some of those killed at the Lattimer massacre (figure 5.4). It reads:

IN MEMORIUM

This Marker Commemorates the Gravesites of Fourteen Unarmed Polish-American Coal Miners Shot to Death As they Marched Peacefully At the Site of the "Lattimer Massacre" Sept. 10. 1897 Seeking to Protect Their Human and Civil Rights.

St. Stanislaus Centennial Committee
Rev. Louis S. Garbacik, Pastor

Figure 5.4. St. Stanislaus Cemetery memorial marker commemorating the gravesites of fourteen of the coal miners killed at the Lattimer massacre. The marker is facing the street. (Photograph by Paul A. Shackel.)

After the centennial event, I paid a visit to the St. Stanislaus Cemetery in 2010 and every year since. I could easily find the graves of the Lattimer martyrs because of the commemorative stone slab, which was visible from the street. However, in early June 2013, repairs occurred to the cemetery wall next to the martyrs, and the stone commemorative marker erected by Fr. Garbacik was no longer visible form the street. I looked on the other side of the wall and found the commemorative marker on the ground, about a foot from the new cinder-block wall, with the inscription facing the wall. Its location made it impossible to read the inscription unless you could squeeze between the wall and the commemorative marker. Several of the grave markers were moved a few feet away from the wall to make room for the commemorative marker (figure 5.5). It is uncertain why the commemorative marker was not replaced on top of the wall and why the commemorative inscription is not visible, even while on the ground. I was wondering if this situation could be an act of defiance by the descendants whose family members once opposed the strikers. The commemorative marker still lies on the ground between the cemetery wall and several of the graves, although its inscription now faces the interior of the cemetery and is readable from a distance.

Several days after viewing the moved marker, I met a faculty member of Wilkes College who grew up in Lattimer. His mother lived on Canal

Figure 5.5. St. Stanislaus Cemetery memorial marker commemorating the gravesites of fourteen of the coal miners killed at the Lattimer massacre. The marker is behind some of the grave markers and facing the cemetery wall, making the inscription difficult to read. (Photograph by Paul A. Shackel.)

Street in Lattimer near one of the archaeological sites we excavated. He noted that for many years the grave markers of the martyrs in St. Stanislaus Cemetery were smeared with mud. When asked why, he noted that there are still many descendants of the posse and the coal police still living around Hazleton. The smearing of mud is a show of defiance against the workers who went on strike and were killed. Fr. Aust is buried in St. Stanislaus Cemetery and has a prominent marker located near the martyrs' burials. On one occasion when I visited the cemetery with students, his gravestone was smeared with mud. I wondered if this, too, could be an act of defiance (figure 5.6).

The anthracite region has a long history of conflict between labor and capital, and it continues today. It is difficult to have only one narrative related to Lattimer. For instance, descendants of Calvin Pardee (owner of the Lattimer Mines during the massacre) wrote a biography of Pardee that includes information about the development of the coal company, as well as the family's perspective of the Lattimer massacre. Foulke and Foulke (1979, 138) wrote that the August 1897 strike occurred not because workers

Figure 5.6. Fr. Aust's grave site in St Stanislaus Cemetery, with mud on his grave marker, January 2010. (Photograph by Paul A. Shackel.)

wanted higher pay, but rather they wanted to secure better treatment for minority groups from eastern Europe. They wrote that John Fahy was sent to the region to organize labor. He succeeded in the area around McAdoo, a few miles outside Hazleton. "They marched in a body on a non-union colliery, intimidating its employees to go on strike. . . . The Luzerne sheriff then swore in 87 deputies, mostly members of the companies' coal and iron police. Each sheriff's deputy was issued a rifle, a revolver and ammunition" (Foulke and Foulke 1979, 138).

Foulke and Foulke (1979, 138–39) continue their pro-industry narrative:

> The sheriff and his deputies met them at the edge of the company property and demanded that they halt. But even though unarmed, they ignored him completely, crossed the property line, and started for the village. Seeing no other way to prevent this unlawful intrusion, which might easily have brought on a fight with company employees, resulting in death or serious bodily injury to many innocent persons, to say nothing of possible property damage, the sheriff ordered his men to stand their ground. After some confusion, a shot rang out, followed by a volley that sounded like a pack of firecrackers. The union men

promptly fled under a hail of bullets. . . . Grandfather and his supporters always referred to the incident as the riot at Lattimer, but union sympathizers called it the Lattimer massacre.

We tried several times to interview descendants of the coal operators and members of the posse. However, they declined to speak with us.

Sometime in the 1970s the roadside marker at the Lattimer memorial was also vandalized. A portion of the text was obscured. The reason for the act is unclear, although on the surface it appears as though it is part of a continuing attack on those supporting labor's heritage (figure 5.7).

During Hazleton's 150th Anniversary celebration in 2007, a play written by local artist Salvador DeFazio recounted the origins of the town. The first part highlights the creation of the town and the leadership provided by its founding father, Ariovistus Pardee. The Pardee story emphasizes his humble beginnings and his development as one of the great coal operators in the region. In the narrative, Ario Pardee passes away in 1892, and at that time a teacher introduces to the audience William Shakespeare's tragedy, *Othello*.

Figure 5.7. Damaged commemorative roadside marker at Lattimer, 1970s. (Courtesy *Standard Speaker*, Hazleton, Pa.)

The introduction of *Othello* in the community provides a parallel story of how the mistrust of foreigners can lead to tragedy. The teacher explained that while the play may seem like a love story, there is a deeper story to tell. It is about the difficulty of maintaining order during a time of social and civic change. Othello is a foreigner and a Moor. He is hired by the Venetian Senate to uphold law and order and a sense of virtue. However, he receives little support when he seeks equality in social matters and marriage (DeFazio 2007). This interpretation of the play makes parallel comparisons to the eastern and southern European immigrant. The play is also a commentary of how the new immigrant is treated in Hazleton today. The narrator explains that—like a perfect storm—misunderstanding, mistreatment, fear, abuse of power, unrealistic demands, and overcapitalization created a revolt. The teacher explains, "Cassio drinks too much and realizes he has put a poison in his mouth, yet, it is Iago who puts poison in the ear of Othello and all ends in tragedy" (DeFazio 2007). The tragedy is the massacre at Lattimer. And the tragedy in *Othello* is also about the misunderstanding of new cultures and customs, as well as different languages, found in the city of Hazleton in 1897.

The area is dealing with almost a century of deindustrialization, and continues along that trajectory. As capital fled the area, and continues to migrate to other regions and other countries, northeastern Pennsylvania's general health and well-being has deteriorated significantly. At the same time, the community is coming to terms with a new episode of immigration. While many feel defeated in this climate of deindustrialization, they have also forgotten the racism their ancestors faced when they arrived a century ago and faced a similar economic situation. Tracing the deindustrialization process in the region provides a framework for understanding the rise of the Tea Party in the 2000s and new forms of racism in the region. DeFazio's 150th anniversary play makes the story of *Othello* relevant in the Hazleton area today.

CHAPTER 6

Deindustrialization and the New Twenty-First-Century Immigrant

Deindustrialization

Deindustrialization, which is the widespread, systematic disinvestment in basic productive capacity, began in northeastern Pennsylvania after World War I. The industrial decline accelerated in the postwar years and remained throughout the twentieth century. With little diversification in the local economies, deindustrialization has severely affected the region (Bluestone and Harrison 1982, 6–9). As a result, the long-term deindustrialization in northeastern Pennsylvania continues to create major demographic shifts. There was an outmigration of heads of households and families as they sought employment elsewhere. More recently, there has been significant population growth in places in northeastern Pennsylvania where new immigrants are seeking low-paying, unskilled jobs in the region's new economy. Much like the commemorative play for Hazleton's 150th anniversary noted in the context of describing Shakespeare's *Othello* (DeFazio 2007), with the new population coming to the region, misunderstanding, mistreatment, fear, and abuse of power are also on the rise.

After World War II, some referred to the United States as the affluent society. However, many parts of Pennsylvania, especially in the anthracite region, lagged far behind. While employment increased 21 percent in the United States from 1947 through 1957, it only increased 4 percent in the anthracite region (Wilson 2003, 184). By the early 1950s, depopulation was evident in the region. While the population in the Hazleton and Wilkes-Barre region declined significantly in the 1950s, and the area did attract some new businesses, the unemployment rate soared to about 18 percent. Mines closed at a rapid pace, and those that stayed open operated only two or three days a week (Dublin 1998, 10).

When widespread unemployment hit, many men left the region and found employment in New Jersey, in steel mills or car assembly plants. Many women found employment in local garment factories. Some started working while their husbands were still employed in the mines, while others started when their husbands lost their jobs. After mining production slowed, and eventually closed, women became the main economic backbone of most households (Dublin 1998, 29). The textile industry in Hazleton endured until the late 1980s, long after peak employment for the coal mining industry in the area. By the early 1990s, Hazleton and most of northeastern Pennsylvania was left without a major industrial center. While some companies closed, the rest moved their machinery to another location with unorganized labor, first moving to the South, then offshore. This strategy is known as "runaway shops." Capital's mobility is a strategy to avoid organized labor. A healthy economy requires perpetual reincarnation, and NEPA (northeastern Pennsylvania) did not attract a major industry after the decline in coal and the exit of the textile industry. Businesses left the main commercial districts and many stores were abandoned and boarded.

Bluestone and Harrison (1982, 12) explain that those who lose their jobs in the old industrial sector rarely have the skills to be employed in the new economy. They write (Bluestone and Harrison 1982, 12), "With industry moving so rapidly, those who lose their jobs in the older sectors of the economy rarely have a chance at employment in the new ones—even within the same region. As a result, the creative destructive process has become synonymous with our conception of the 'throwaway' culture. Instead of recycling people and communities through the development process, the pace of capital mobility has become so fast that people and communities are carelessly discarded to make room for new ones." They describe the fundamental struggle between capital and community and quote from planning theorist John Friedman. "The capitalist city has no reverence for life. It bulldozes over neighborhoods to make way for business. It abandons entire regions, because profits are greater somewhere else. Deprived of their life spaces, people's lives are reduced to a purely economic dimension as workers and consumers—so long, at least, as there is work" (Freidman quoted in Bluestone and Harrison 1982, 47). Instead of providing new employment opportunities, a higher standard of living, and enhanced security, the decisions of managers are doing just the opposite.

The 1967 *Hazle Township Comprehensive Plan* (Kendree and Shepherd 1967a, 1967b) describes the impact of deindustrialization in the region. In the mid-1960s, housing stock in the area was considerably older than the rest of the state. In the mid-1960s, the Commonwealth of Pennsylvania had 64 percent of its housing built before World War II, compared to 89 percent in Luzerne County. In the Hazle Township 9 percent of dwellings were considered "deteriorating" and 4 percent were deemed "dilapidated" (1967a, 1–7, 3–14).

Though it is not actively mined today, the landscape is dominated by strip mines and mine dumps. There are vast amounts of loose rock materials, deep trenches, and high cliffs. While stripped land dominates the landscape, there are also a number of breaker waste deposits. Mine dumps contain coal and are potential fire hazards. Because there was no organized trash collection system, households disposed of their trash in backyards, mine dumps in abandoned strip mines, or along abandoned stripping roads. Air pollution was a concern, from mining, industry, and occasion fires of culm banks, where trash was stored and ignited. Many junkyards in the area were not regulated (Kendree and Shepherd 1967a, 319–21). Raw sewage was dumped into creeks, which drained into the Susquehanna River. These unregulated sewer systems were called wildcat sewers (Kendree and Shepherd 1967b, 1–6).

In 1967, the Hazle Township planning document described the desolate and destroyed landscape. "To the casual visitor the strippings themselves represent a fantastic impression of the amount of labor that has gone into the creation of these sites. The distortion of the landscape is gigantic in its proportions, as it is the largest concentration of man-disturbed terrain in the world. Large spoil banks and the high walls of the strip pits provide excellent vantage points of this destruction. Even more impressive is the view from the interior of a strip pit. In many such locations there is no sign of any life whatsoever that can be seen" (Kendree and Shepherd 1967b, 5–2).

Deasy and Griess (1961, 1–8) wrote in 1961 about the harsh realities of high unemployment in the anthracite region. They proposed diversification of the region's economy by building the area's tourism industry. However, to move in this direction, a new and improved infrastructure was needed, like the construction and improvement of roads and tourist service facilities, as well as advertising and promotion. They marveled that the anthracite region had so much earth-moving equipment in such a limited area. Deasy and Griess believed that tourists would enjoy seeing the power of the machinery extracting coal, as well as the power of the towering breakers, that this would have "an almost hypnotic appeal" (Deasy and Griess 1961, 4). They also thought tourists would find of interest the small-scale one- or two-man operation where visitors could "strike up an acquaintance with a real miner" (Deasy and Griess 1961, 4). Geological features and strata exposed by strip mining could also draw tourists. In particular, the shale being extracted along with the coal could also provide an opportunity for tourists to find fossilized plants. Presenting these as fossil hunting grounds might draw interest. In 1961, the only facility catering to outside visitors was a small anthracite gift shop in Frackville, and the City of Scranton operated a mine for tourists (Deasy and Griess 1961, 4).

Hazleton had a population of thirty-eight thousand during World War II and declined to about twenty-three thousand by 2000. Within five years the city's population grew to thirty-one thousand, and it now hovers around twenty-five thousand residents. Other surrounding towns have felt the impact of the declining coal industry and the flight of its population. For instance, Shenandoah, which once had a population of twenty-five thousand, now has about five thousand people. Mahonoy City, which once numbered fifteen thousand, now has about four thousand residents (Bahadur 2006; Englund 2007, 887; Tarone 2004, 128).

Hazleton's decline is more complicated than simply industry leaving the area. The economic blight of the region has its foundation in the conscious decisions of coal barons to not encourage higher education. When the coal barons offered philanthropy, they helped establish colleges outside their region. They wanted an uneducated population because they were worth less in the job market. So, while the cities of Wilkes-Barre and Scranton eventually worked to establish colleges, not until 1934 did the Penn State campus at Hazleton open. Because of the lack of higher-education opportunities in the area, many technical industries did not look to Hazleton as fertile ground for establishing their high-tech business and manufacturing. The lack of opportunities also created a "youth flight," a phenomenon that continues today (Tarone 2004, 5). According to the 2012 census, 26.7 percent of Pennsylvania's population has earned an undergraduate degree or higher. The average of the counties in the anthracite region is about half that of the state (Keil and Keil 2015, 119).

An anti-education sentiment persists in the community, which I found on a local cable television program that focuses on community concerns—the *Sam Lesante Show*. From YouTube I downloaded the episode titled "Political Watchdog Groups with Louis 'Booty' Beltrami" (SSPTV.com http://ssptv.com/new/political-watchdog-group-with-louis-booty-beltrami/). Louis Beltrami is a former coal operator and now a restaurant owner. He is a major proponent of the anti-immigration movement in Hazleton, as well as being anti-union. In Beltrami's interview with Sam Lesante, he spoke about the local community's fiscal crisis. In a few anti-union statements, he noted that schoolteachers, police officers, and firefighters enjoy too many benefits, which the community should not be funding. Beltrami viewed collective bargaining as an action against neighbors. He also said that we should not aspire to send everyone to college. After all, he explained, who is going to grease your car and perform all of the blue-collar work needed for the community to operate efficiently?

The loss of high-paying manufacturing jobs in the region has placed downward pressure on the average wage in most sectors, thereby lowering the cost of labor. The decline in manufacturing also weakened the unions, and thereby made it easier to make changes in the political system to benefit corporations and individual investors (Keil and Keil 2015, 115).

In the twenty-first century, the anthracite region continues to lag behind in almost every economic indicator in Pennsylvania. For instance, the average weekly wage is about 20–25 percent lower in the region when compared to the rest of the state. In 2012, the poverty rate for Pennsylvania was 13.2 percent, while in several counties in the anthracite region the rate stood well above 16 percent (Keil and Keil 2015, 91).

The Brookings Institute report, "America's Advanced Industries," rated the top one hundred metropolitan areas for advanced or STEM-related jobs (science, technology, engineering, and math). The Scranton/Wilkes-Barre/Hazleton MSA (Metropolitan Statistical Areas) ranked 92nd.These jobs are leaving the area at a rate of about 2 percent per year, and the area has about half the number of advanced industrial jobs compared with 1980. Salaries lag significantly when compared to other regions. The report indicates that advanced industries are essential for economic renewal in the United States (Falchek 2015; Muro et al. 2015).

Economists compared 367 MSAs in the United States in a study that considered overall happiness. Northeastern Pennsylvania (Scranton/Wilkes-Barre/Hazleton MSA) ranked last (Glaeser, Gottlieb, and Ziv 2014). In addition, the area has one of the lowest rates of population attaining higher education, and the quality of primary education ranks as one of the lowest in the country (Bernardo 2016). The offshoring of manufacturing jobs has led to the deskilling of the area's workforce and the devaluing of the economy, making it one of the most undesirable places to relocate manufacturing. In 2015, northeastern Pennsylvania was rated the third worst place for small businesses among the nation's one hundred largest MSAs, according to WalletHub.com. This organization claims to be a "social network for your wallet." The study incorporated data from the U.S. Census Bureau and Bureau of Labor Statistics, the Kaiser Foundation, and the United States Conference of Mayors, and analyzed businesses with fewer than 250 employees. The survey examined employee health insurance, employee earnings, and average hours worked, among other variables. For the Scranton/Wilkes-Barre/Hazleton MSA, the region also ranked poorly in 1) median annual income

($49,954, 65th among 100 MSAs), 2) unemployment rate (7 percent, 88th place), 3) well-being index (84th place), 4) average number of hours worked by small-business employees (35.5, 84th place), and 5) projected population growth to 2042 (2.6 percent, 92nd place) (Guydish 2015; Keirnan 2015).

Along with the economic stress in the region, the community is also distressed in terms of physical and emotional well-being. In 2014 a report, the *State of American Well-Being,* was released by Gallup and Healthways regarding Americans' emotional health, work environment, physical health, healthy behaviors, and basic access to health care. About 178,000 people across the United States were interviewed, including a sample size of 1,092 from northeastern Pennsylvania. The survey ranked northeastern Pennsylvania 177th out of 189 metropolitan areas with regard to general well-being (Gallup-Healthways Well-Being Index 2014). Regional newspapers, like *The Citizen's Voice* (Halpin 2014), reported, "Region's residents among most miserable." Experts in the region indicate that the results are probably related to the continuing economic downturn. The director of the social work program at Misericordia University, Margaret Rapp, described the survey results of the community being afraid of change, "because they've been burned so many times resulting in this almost fear cycle" (Halpin 2014).

Recent studies by the Center for Disease Control and Prevention show a high rate of deaths related to heart disease in the American South among the white population (Steckel and Senny 2015), and this high incidence rate reaches into northern Appalachia, which includes the anthracite coal region of northeastern Pennsylvania. The area has a high mortality rate when compared to the rest of the state (Keil and Keil 2015, 117).

Deteriorating economic conditions as a result of deindustrialization and the persistence of high rates of chronic disease in northeastern Pennsylvania are likely a product of long-term inequities rooted in the racial attitudes the coal barons reinforced while recruiting, employing, and exploiting new immigrants from eastern and southern Europe beginning in the late nineteenth century. Keil and Keil (2015, 118) note that "In sum, what the anthracite industry has left behind is a befouled landscape in most of the counties in which the industry operated: a locale that has been unable to sustain high levels of economic growth; a population that has significant health problems and a profile of higher-than-average health risks; and a region where one has little hope for a prosperous and healthy future."

There have been several efforts to address the faltering economy. The best-known is CAN DO (Community Area New Development Organization). The organization developed in the mid-1950s as a private, nonprofit industrial/economic development corporation serving Greater Hazleton. The mission is to improve the quality of life in the Greater Hazleton area through the creation and retention of a full range of employment opportunities.

While CAN DO has developed lands and attracted businesses to the area, in the late 1990s there was a statewide strategy to create economic development stimulation projects to help revitalize economically distressed areas. In 1998, Governor Tom Ridge signed Keystone Opportunity Zone (KOZ) and in 2000, the legislation was amended with the passage of Keystone Opportunity Expansion Zone (KOEZ) and Technical Change Bill (Act 119 of 2000). The program was further expanded in 2002, which created the Keystone Opportunity Improvement Zone (KOIZ). The KOZ/KOEZ programs provided tax abatements to businesses. KOZ serves as a magnet for "big box" warehouses, fulfillment centers, and a meatpacking plant. The majority of these jobs are temporary, low paying, and sometimes dangerous. Many of these companies in KOZ/KOEZ have the reputation for exploiting immigrant labor (Longazel 2016, 19).

Though jobs were created with the KOZ/KOEZ initiative in Pennsylvania as a whole, and in the anthracite region in particular, the state has had a net loss of high-wage manufacturing jobs since 2000. In fact, even before the Great Recession of 2008, between 2000 and 2007 Pennsylvania lost more than 207,000 manufacturing jobs, or about 25 percent of its manufacturing base. These jobs have been replaced by lower-paying jobs, in health care and social assistance, where the pay is about 30 percent lower. The northeastern Pennsylvania region lost more than twelve thousand manufacturing jobs, or about 28 percent of its manufacturing base (Keil and Keil 2015, 100–102).

In recent years, there has been a new pattern of immigration into the United States. New immigrants are not only going to the traditional gateway communities, they are also settling in the country's small interior towns (Longazel 2016, 7). The KOZ/KOEZ program has attracted companies like Office Depot, Amazon, American Eagle Outfitters, Michael's craft store, Wegmans, and a Cargill meat-packing plant (Jackson 2015). Since the early 2000s, the Hazleton area has received an influx of immigrants from South America and the Caribbean who are filling these jobs in the KOZ/KOEZ program. For instance, one new industry in the KOZ/KOEZ program is Cargill, which employs about six hundred workers. In a town of about twenty-five thousand residents (2010 U.S. Federal Census), that is a large part of the labor force.

About 70 percent of workers in the KOZ/KOEZ program are Latina/o migrants from the Dominican Republic, Mexico, or Peru. Many of the new immigrants first entered through New York and New Jersey, and they came to Hazleton because of the low cost of housing and for work opportunities. By 2006, the city had grown to thirty-one thousand people, with 30 percent of its residents identified as Latina/o. In 2009, Luzerne County ranked number one in the United States in Latina/o population growth (Longazel 2016, 20). A recent Pew Report lists Luzerne County as one of the top ten counties in Latino population growth from 2000 to 2011, at 538 percent. Other population growth centers are in non-border states like Georgia, South Dakota, Mississippi, Wisconsin, Virginia, and North Carolina (Brown and Lopez 2013, 12). The 2010 U.S. Federal Census indicates that the town consisted of more than 37 percent Latina/o residents, and in 2017 it approached 50 percent.

Many of the white citizens of northeastern Pennsylvania interviewed over the past several decades described themselves as self-made men and women who achieved a standard of living because of their hard work and struggle. Several of those interviewed expressed a disdain for the newcomers to the region—African Americans and Latinos. They see their traditional life being threatened by the newcomers, and they see the newcomers relying excessively on welfare programs. They do not see the similarity between their immigrant history of poverty and survival and the situation of their contemporary new immigrant neighbors (Dublin 1998, 31).

Illegal Immigration Relief Act

At Hazleton city council meetings, people often express their anger at the new immigrant by linking them with the city's quick decline. This sentiment generally ignores the many decades of decline since the near-total collapse of the coal industry after World War II (Longazel 2016, 41). Longazel (2016, 81) discusses the politics of divide and conquer in his book *Undocumented Fears: Immigration and the Politics of Divide and Conquer in Hazleton, Pennsylvania*. He notes that the negative sentiment toward the new immigrant, which is predominantly Latina/o, is not necessarily about race hatred, but rather about using race to create divisions. It is about creating distinctions that make people appear fundamentally different from those who are affirmed and granted inclusion. Binaries are created within groups, in this case with regard to citizenship—legal vs. illegal. A few Hazleton restaurant had signs in their windows: "Illegals not served here," reminiscent of a similar phrase the Irish faced in the mid-nineteenth century.

This sentiment for a pre-Latina/o city is reinforced by the city newspaper. For several years the city newspaper, the *Standard Speaker*, ran a column with people reminiscing about Hazleton before the new immigration. There were stories about shopping downtown, ice cream shops, and the drive-in theater. The newspaper was actively reinforcing the community's ideal past they believed changed with the new immigration. This column helped maintain and reinforce the traditional narrative at the expense of the new immigrants.

It is difficult to ignore the presence of the new immigrant. For instance, the area once known as Donegal Hill, named after the county in Ireland from where many of the original residents emigrated, has changed dramatically. St. Gabriel's Church, which is located on Wyoming Street in Donegal Hill, has a lively noontime Spanish-language Mass on Sundays. Shops along the street cater to the new Spanish-speaking residents. More recently, Broad Street, which was boarded up and only had a few of the older shops and restaurants remaining, is now revitalizing with Latino restaurants, botanicas, and clothing stores, all catering to the new Latino population. The presence of the Latino community is evident. At the same time, the traditional population is feeling crowded and overwhelmed.

Leo Chavez has examined the negative sentiment the new Latina/o immigrants are facing in what he terms as the Latino Threat Narrative. He notes that this narrative describes the new Latino immigrant "is part of an invading force from south of the border that is bent on reconquering land that was formerly theirs (the U.S. Southwest) and destroying the American way of life" (Chavez 2008, 2). "Akin to what some scholars refer to as a racial bribe, the Latino Threat Narrative in this way promises a symbolic uplift to those White workers who choose to ally with White political and economic elites instead of conspiring with their fellow workers of color to protest their shared economic plight" (Longazel 2016, 7). The new immigrants are seen as takers. For instance, one City of Hazleton council member stated that the undocumented immigrant use the emergency room as their primary care. They go to the emergency room for colds, coughs, and splinters (Longazel 2016, 36). "Once officials mold this already familiar narrative to fit the local context and to account for particular local conditions and events, the story becomes one of 'greedy,' 'crime-prone,' 'outsiders,' coming in uninvited to disrupt an otherwise 'tranquil' 'small town' lifestyle" (Longazel 2016, 7).

While it was the mid- to late-nineteenth-century immigration from Europe that fueled the growth of the region, it seems ironic that the anti-immigrant movement is being promoted by the mayor (now U.S. congressman), Lou Barletta, a descendant of Italian immigrants. Frustrated by Congress's failure to develop new immigration policy in the early 2000s, many other municipalities began legislating immigration laws and policies (Vicino 2013). In July 2006, Mayor Barletta and the city council passed an ordinance titled "Illegal Immigration Relief Act" because they felt the federal government was not doing its job to control illegal immigration. The act asserted, "Illegal immigration leads to higher crime rates, contributes to overcrowded classrooms and failing schools . . . and burdens public services." The goal of the ordinance was to "abate the nuisance of illegal immigration by diligently prohibiting the acts and policies that facilitate [it]" (quoted in Englund 2007, 888). Hazleton used similar legislation proposed by the city of San Bernardino, California. The author of the San Bernardino legislation claimed the immigration legislation would save "California from turning into a 'Third World cesspool' of illegal immigrants" (quoted in Jordan 2006, 5B). When the city council passed the ordinance, it placed Hazleton at the center of the U.S. immigration controversy, even though it was several thousand miles from the U.S.–Mexico border (Longazel 2016, xii).

One of the catalysts for the development of Hazleton's anti-immigration legislation was the perceived dramatic increase in the homicide rate. The Uniform Crime Reports from January 2000 to May 2006 list only four homicides in Hazleton. However, when two undocumented Latinos, Pedro Cabrera and Joan Romero, were charged with the killing of a white Hazletonian, the case received much more attention from the city police and from city officials than previous homicide cases. The *Standard Speaker* began using terms like "illegal immigrants," rather than the less-loaded "undocumented immigrants." The newspaper reported the men lived together and "came to the United States illegally from the Dominican Republic." Hazleton was seen as being under siege and needing to get tough on crime. Officials began to look for some type of ordinance to protect their city (Longazel 2016, 27–29). The Latino Threat Narrative was operationalized even though prosecutors later dropped the case from lack of evidence (Longazel 2016, 28).

The Illegal Immigration Relief Act of Hazleton punished businesses if they hired an illegal immigrant by suspending the business owner's license for five years for the first violation and ten years for the next. However, the ordinance did not provide any mechanism for identifying an immigrant's status. The ordinance also required all renters in Hazleton prove their citizenship at City Hall, and it prohibited landlords from knowingly renting or leasing their property to illegal aliens. A city imposed a fine of $1,000 per day for renting to an illegal immigrant. The ordinance did not establish a definition for the term "illegal immigrant." It also established English at Hazleton's official language. The contested legislation has divided the community and has made both legal and illegal immigrants feel unwanted (Bahadur 2006, Englund 2007, 884; Kroft 2006). Over the course of eight months in 2006 and 2007, Hazleton enacted—and amended—a series of ordinances designed to make it more difficult for undocumented immigrants to live and work in the city (Chishti and Bergeron 2014). The penalties for business owners and landlords became less harsh, and the term "illegal alien" was now defined by United States code; however, the intention of the ordinance remained the same. Some towns like Altoona, Pennsylvania, have followed Hazleton's lead, even though they do not have a problem with immigration (Englund 2007, 884). Some critics of Barletta's ordinance see it as an affront to the city's immigrant heritage (Englund 2007, 887).

Hazleton's development of the Illegal Immigration Relief Act occurred during the 2006/2007 U.S. congressional debate over immigration reform. After long debates, Congress was unable to address the status of about twelve million undocumented immigrants. This inaction provided the stimulus for many lawmakers on the state and local levels to act. Throughout the United States, protest marches attracted hundreds of thousands of supporters demanding comprehensive immigration reform. These demonstrations were countered by communities experiencing rapid immigration growth, which, in turn, galvanized a movement across the United States (Chishti and Bergeron 2014). Between July 2006 and July 2007, more than one hundred immigration enforcement proposals were under consideration across the country. During the new millennium's first decade, 107 U.S. towns, cities, and counties had approved local immigration enforcement ordinances (Chishti and Bergeron 2014).

In 2006, the Hazleton immigration issue received national attention. A CBS broadcast of *60 Minutes* featured Mayor Lou Barletta, Senator Rick Santorum, and Chris Simcox of the Minutemen Civil Defense Corp. They rallied in Hazleton against the influx of new Latino immigrants. Mayor Barletta claimed that illegal immigrants were overwhelming the city's resources and ruining its citizens' quality of life (Kroft 2006). He asserted that the police force was too small for the city of its size, and they were responding to more serious urban crimes, like drive-by shootings and the sale of illegal drugs on playgrounds. The un-reimbursed medical expenses for emergency room visits were up by 60 percent, and public school enrollment had increased by 25 percent. The budget for teaching ESL (English as a second language) grew from $500 per year to more than $875,000. How these changing numbers were related to illegal immigrants is unknown, and the mayor himself did not have a good sense of how many undocumented immigrants were in Hazleton (Kroft 2006; Englund 2007, 887).

When asked to provide data linking the higher crime rate with undocumented immigrants, the mayor responded: "I don't need a number. . . . Numbers are important mostly to people from the outside who are trying to understand what's happening. But if you lived in the city of Hazleton and you woke up to morning news such as this [referring to the crimes], you would understand that we have a major immigration problem" (Barletta quoted in Jackson 2006, b5).

From the time immigration surged in Hazleton, from 2001 to 2006, a perception existed that there was an accompanying surge in crime (Longazel 2016, 36). The *Philadelphia Inquirer* noted that the police chief made statements about illegal immigrants without any firm data. He told the newspaper he did not know how many illegal immigrants lived, worked, or went to school in Hazleton. The chief was not even certain about how many of the crimes in Hazleton were committed by illegal versus legal immigrants. In reality, the statistics show a surprising contradiction in the mayor's impression of crime in Hazleton. While the town's population soared to thirty-one thousand in 2005, an increase of about eight thousand residents in five years, theft and drug-related crimes rose from eighty incidents in 2001 to 127 in 2005, according to the Pennsylvania Uniform Crime Reporting System. Other crimes like rape, robbery, homicide, and assault decreased. Arrests in Hazleton dropped from 1,458 in 2000 to 1,263 in 2005 (Bahadur 2006). It had a higher violent crime rate versus the national average only in 2010 and 2011. In no years between 1999 and 2011 did the city have a property crime rate that exceeded the national average (Keil and Keil 2015, 123).

Despite negative reaction toward the recent arrivals, statistics show that the new immigration has had a clear economic benefit to the City of Hazleton. While the population has increased significantly, filling empty housing stock throughout the depressed city, the city budget now shows a surplus, rather than the $1.2 million deficit in 2000. Many new businesses opened up in Hazleton, and home values increased 125 percent. In 2005, Mayor Barletta claimed Hazleton had reached its "healthiest state in decades" (Englund 2007, 888). Therefore, the line in the sand appeared to be clearly drawn. Those supporting the rights of immigrants point to the positive economic growth, and those opposed to immigration claim the new immigrant is a drain on local social services.

The Reaction to the Illegal Immigration Relief Act

Many of the Spanish-speaking business owners have felt the surge of xenophobia after the passing of the act. "'We feel everything change. Non-Latinos look at us, and they think we're illegal. Never before this happens, that they say, Go back to your country,' said one documented resident from Columbia. 'You can see contempt in people's faces. You can see the rejection,' noted another Latino resident" (quoted in Bahadur 2006).

After a *CBS Evening News* broadcast (March 13, 2007) covering the immigration showdown in Hazleton, many viewers responded to the broadcast's transcripts, and bloggers posted their response. By a margin close to ten to one, the opinion of bloggers was negative toward the case of the city's undocumented residents. Following are a few of the reactions. One blogger equated those sympathetic to the undocumented worker to anarchists. "I cannot understand why anyone, other than anarchists, should oppose the use of the Rule of Law as one of the cornerstones of any civilized country" (laurairby, March 14, 2007, 10:15 P.M. EDT). Another blogger was not pleased the ACLU was involved in the case. "The ACLU should be fined for aiding and abetting criminals" (olebd, March 14, 2007, 9:50 P.M. EDT). Several bloggers used an economic argument to criticize the undocumented workers. "The illegal not the legal are draining our country just like a knife to the throat drains the life out of a living animal" (by frankbowers, March 14, 2007, 9:35 P.M. EDT). Another stated, "Glad to see someone finally has the gonads to start squelching these leeches on the taxpayers. High fives to Hazelton" (mcjohn2, March 14, 2007, 9:19 P.M. EDT). Some used the lack of assimilation as a way to justify deportation. "If they want to come here, they need to learn to read, write, and speak English first. Then pay their way to get here. If they don't do this, don't attempt it. America is no stomping ground for every riff-raff that wants to come here. It disgusts me every time I walk down the street and I try to talk to one of these illegals and they can't even speak English. This is a downright shame. We have given this country away and we need to get it back" (by hhusted, March 14, 2007, 1:55 P.M. EDT). Taking this country back has led to some unsympathetic expressions and even a call for violence. One blogger states, "They need to build a huge wall along the border between the United States and Mexico. Have the military guard that border with orders to shoot to kill any illegal who enters our country, then after their [*sic*] dead, throw their worthless body back into Mexico as a warning to others there that this will be your fate if you dare enter our country in this way" (noaanhc, March 14, 2007, 10:57 A.M. EDT).

Some of my own friends and acquaintances have related stories of being a new immigrant in Hazleton. During the summers, we lived in a predominantly Latino neighborhood. Our Latino neighbors moved to Hazleton several years ago. One reported he is an ex-police officer from Los Angeles, and moved to Hazleton from New Jersey because of the low cost of housing. He said he needs to save his money to send his seven daughters to college. My neighbor purchased his modest house for about $40,000 (interview, June 8, 2014).

Our neighbor told us a story about when he first moved to Hazleton. He was driving in the city with a new Pennsylvania license plate on his car, but he still had a New Jersey driver's license. I am not sure why he was pulled over by a Hazleton City police officer. He described the police officer about twenty years old. The officer asked him for his documents. He lowered his window and gave him his license, registration, and insurance card. The officer once again asked for his documentation. Our neighbor replied that he gave him the appropriate information. As an ex-police officer from Los Angeles, he said he knew nothing further was required of him. When the officer asked him again, he raised his window and called the state police. According to our neighbor, when the state trooper came, he scolded the Hazleton police officer, telling him that it was not legal or appropriate to ask for additional information (interview, June 8, 2014).

With a city that has a Hispanic population approaching 50 percent, there is only one Spanish-speaking police officer on the police force, which makes it difficult to intervene, converse, and communicate during citizen interactions. A large portion of the Hispanic community feel the police force is prejudiced against them, and they do not feel they will receive a fair shake in any interaction. We witnessed one incident where a motorcycle accident led to a quick and efficient response from the police. They wrote and filed a report regarding the accident. Surprisingly, they did not call an ambulance for the injured Latino who was sprawled out on the concrete sidewalk. The police left, and the man called a friend, who drove him to the hospital. A few days later, his family took him to a New York City hospital to get his ankle bones reset.

We interviewed several people in the greater Hazleton area, with the hope of soliciting their perspective about the treatment of the historic immigrants, the killing of innocent men at Lattimer, and the treatment of the new immigrants today. Some made the connection between the historic and contemporary immigrant. For instance, Bobby, in his early twenties, is a lifelong resident of the greater Hazleton area. He described Lattimer in terms of its historic significance within America—as part of the Industrial Revolution, a time when good men came to America in search of better lives for themselves and their families. He explained: "Back then if you didn't adhere to what [the mine operators] were saying, there were some pretty dire consequences to pay, if you didn't go along with them. In my opinion they were, a lot of the miners back then—not necessarily these massacre victims—but, a lot of these miners were pioneers in a sense, and they were . . . heroes in a lot of ways" (interview, March 20, 2010).

When asked about how the massacre might be important today, he saw very little connection, other than that the massacre was significant within the history of coal mining, an industry that shaped the social and economic fabric of the region. He said that those immigrants are nothing like the immigrants now, who are in search of a cheaper way of life:

> A good portion of the immigrants that came off of Ellis Island . . . were coming because they were hard workers, but their situation, wherever they were coming from, wasn't the best; and they knew that America was the promised land, and they could make a new start. Whereas today I think a lot of people come here because they know how easy it is to live off of the government. . . . A lot of them you see are fairly lazy. . . . They're not trying to get their papers like [the] Italians or Germans or Irish did. They're coming here because . . . someone said, "Hey, you could live in Hazleton for very, very cheap prices and you don't even have to tell them that you're there." . . . The big difference [between 1897 and now] is why they were immigrating then as compared to now (interview, March 20, 2010).

Many of the descendants in the anthracite region have forgotten their immigrant roots and the resolve of their ancestors to petition for better living wages and working conditions. A little over a century ago, scholars described the new immigrant in derogatory terms. For instance, Peter Roberts (1970 [1904], 11) wrote, "The Sclavs represent possibly the lowest grade of European workmen that can be imported. . . ." In the Lattimer court case, the defense attorney used the foreign scare for his case against the non–"English speaking" workers. Palmer explained, "In four years this country will be all Hungarian. . . . Is the barbarism from the East to descend upon us and wipe us off the face of the earth" (*The Press,* March 5, 1898). Substitute Latino for Sclav or Hungarian in the above statement, and that appears to be the sentiment of many in Hazleton today regarding the new immigrant.

The Ordinance

The Illegal Immigration Relief Act was to go into effect November 1, 2006; however, the ACLU and other civil rights groups brought litigation. They argued that the ordinances illegally usurped the federal government's exclusive power over immigration. It also violated due process and equal protection rights (American Civil Liberties Union 2015). The ACLU noted that:

> If the ordinance is allowed to stand, anyone who looks or sounds foreign—regardless of their actual immigration status—will not be able to participate

> meaningfully in life in Hazleton, returning to the days when discriminatory laws forbade certain classes of people from owning land, running businesses or living in certain places. (quoted in McKanders 2007, 12)

The ACLU also explained:

> Many of those affected by the overly broad ordinance are here legally and have lived, worked and worshiped in Hazleton for a long time. In desperation and fear, some of those residents have already decided to close their businesses, move out of Hazleton, or, simply hide as best as they can behind closed doors. (quoted in McKanders 2007, 12)

A court issued a restraining order, which suspended this implementation in November 2006. Two weeks later, the federal judge extended the order by 120 days. In March 2007, the case came before Judge James Munley of the Middle District Court in Scranton, Pennsylvania, to assess *Lozano v. Hazleton*. The plaintiff was led by a coalition that included the ACLU, Puerto Rican Legal Defense and Education Fund, the U.S. Chamber of Commerce, and the U.S. Council of Catholic Bishops. The ACLU lawyers pointed out that the new economic growth in Hazleton owed to the new immigrant. New settlers established roots in the community, paid property taxes, and worked hard. While the city claimed crime was up, the ACLU lawyers pointed out that the city lacked data to connect this trend to the new immigrant. On the other hand, the nativist bloggers asserted that these statistics were useless for making the connection to a specific ethnic group because of "proponents of illegal aliens like the ACLU who have made sure that these statistics cannot be collected by hospitals, schools and prisons." There were also major drug busts involving the gangs MS-13 and the Latin Kings (Digger, March 13, 2007; Longazel 2016, 52).

On July 26, 2007, the District Court for the Middle District of Pennsylvania ruled the Illegal Immigration Relief Act unconstitutional and prohibited Hazleton from enforcing the associated laws (Baratta, July 26 2007). The judge wrote a 207-page decision and stated that immigration comes under the jurisdiction of the federal government. This decision stems from an 1875 the Supreme Court ruling stating that the power to regulate immigration belonged exclusively to the federal government. The ruling is the outcome of New York City passing a law requiring every foreigner arriving in the city to pay a bond to ensure that newcomers would not become a future charge to the state. The Supreme Court found that immigration was a commerce issue. Commerce includes the vessels entering the country, which also consists of a vessel's cargo and passengers. Because the new immigrants contributed to the country's manufacturing and agriculture, the Court held that immigration lay within the commerce power given to Congress to regulate. Because the power to regulate commerce lies exclusively within the authority of Congress, therefore the right to regulate immigration is derived from this, and must also be exclusively under the jurisdiction of Congress (Englund 2007, 891–93; McKenders 2007, 14). While the decision by the City of Hazleton to prohibit the employment of illegal immigrants does not involve a decision about who may enter the country, such a law can be viewed as an effort to control immigration (Englund 2007, 898–99).

The law also violates the rights the Constitution guarantees to every person in the United States, whether a documented resident or not. In regard to landlords renting to undocumented immigrants, the judge referred to the Civil Rights Act of 1866. It states that all persons within the jurisdiction of the United States shall have the same rights in every state and territory, "to make and enforce contracts, to sue, be parties, give evidence, and to the full and equal benefit of all laws and proceedings for the security of persons and property as is enjoyed by white citizens, and shall be subject to like punishment, pains, penal-ties, taxes, licenses, and exactions of every kind, and to no other" (Englund 2007, 899). In other words, the act makes it illegal to discriminate in jobs and housing based on the classification of race. This type of discrimination, based on race, color, or national origin, is also prohibited by the Civil Rights Act of 1964.

In general, once an immigrant has entered the United States lawfully, he or she is free to move from state to state and jurisdiction to jurisdiction. Municipalities cannot create laws to restrict their movement or justify ordinances on the grounds that the new immigrant is depleting local resources.

Even though the ordinance was found unconstitutional, it had its desired effects. Many immigrants moved out of town when the legislation was first proposed (Jackson 2006, B5). What the Illegal Immigration Relief Act effectively accomplished was to associate undocumented immigrants with gangs, drugs, and rising crime. Many of the legal immigrants felt harassed and moved to other municipalities. Statistics related to crime in Hazleton could not be clearly associated with the undocumented immigrant and were, at times, exaggerated by city officials. Some residents refused to be interviewed for their response to the court case because they feared some form of retaliation. Others of Latin descent spoke about how the failed ordinance had begun to dissolve their community, as hundreds of people have left the municipality. One resident, a naturalized citizen from the Dominican Republic, explained how employers use undocumented immigrants to their advantage. They work for lower wages and drive down the wage scale, forcing legal residents to accept abusive working conditions (Gass, July 27, 2007).

Posted on David Duke's Web page is the transcript from the Lou Dobbs show that highlights the immigration issue in Hazleton. Duke, a white supremacist who lives in Louisiana, is also a former Grand Wizard of the Knights of the Ku Klux Klan. He is a former Louisiana state representative, and he has run for state and federal office promoting a white supremacist platform. Duke rekindled the National Association for the Advancement of White People in 1980 as a way to counter the efforts of the NAACP (Duke 2007, July).

Mayor Barletta responded on national television by stating that the judgment was "an injustice not only to the city, but to those around the country." He claimed that Hazleton was going to appeal to the Third Circuit Court of Philadelphia. The mayor then exclaimed, "I'm prepared to fight this all the way to the Supreme Court. Today was a slip and not a fall, and this battle is far from over" (quoted in Dobbs, July 27, 2007). Mayor Barletta explained that he thought it was "amusing" that the judge ruled the city of Hazleton could not enact laws addressing actions the federal government should be handling. "If they were doing their job, obviously, I wouldn't have to take a stand" (Barletta quoted in Dobbs, July 27, 2007).

The Third Circuit Court of Appeals upheld the ruling in 2010. However, in 2011, the ordinance was resurrected following a U.S. Supreme Court decision affirming a facet of Arizona's anti-immigration law punishing employers who hire undocumented immigrants. The case went back before the Third Circuit appeals court after the Supreme Court's 2012 ruling on Arizona's anti-immigrant laws. In July 2013, the Third Circuit Court of Appeals reaffirmed its earlier finding. That decision came the same week the Fifth Court of Appeals ruled against a similar anti-immigrant ordinance in Farmers Branch, Texas (American Civil Liberties Union 2015).

The City of Hazleton brought the case to the Supreme Court and, in 2014, the Court denied Hazleton's request that it hear the case, thereby leaving in place the federal court's ruling, which found the ordinance unconstitutional. "The decision by both circuits to strike down the rental ordinances as unconstitutional—and the Supreme Court's recent decision not to review these cases—will make it hard for future similar housing ordinances to pass judicial scrutiny" (Chishti and Bergeron 2014). Plaintiffs' petition for approximately $2.8 million in legal fees is currently pending in the district court (American Civil Liberties Union 2015).*

* For a good overview of the events related to the Illegal Immigration Relief Act, visit the ACLU Web page, https://www.aclu.org/search/hazleton.

Much of this fear and the attempted xenophobic legislation we see today in NEPA are reminiscent of the 1897 Campbell Act (aka Alien Tax Law). The Commonwealth of Pennsylvania enacted a law contributing to the frustrations that resulted in the Lattimer massacre. Though the Campbell Act (and the Illegal Immigration Relief Act) was deemed unconstitutional, some of the company owners, as well as an editorial in the *Philadelphia Times,* supported the act and insisted that its constitutionality come before the Supreme Court. They claimed the new immigrant did not pay his or her fair share of taxes. The *Hazleton Sentinel* (August 31, 1897b), among other newspapers, responded by noting that these men did pay their fair share. They were severely underpaid. These men did pay rent to their landlords, who then paid taxes from the proceeds. The role of the government, noted the editorial, is not to squeeze resources from one group of people for the advantage of another:

> The conditions in which the foreigner has lived bordered on servitude. He has been badly treated and badly paid. . . . Now that this class sees they have been used as a medium for combatting the intelligence of natives in putting down the strikers of old, decide to make demands considered fair before they are denounced as savages and coolies who have no right to complain. They might not be successful just at present but their grievances will not be aired again, and it is but a question of time when reasonable accessions will have to be made. They are now in the majority and are not slow to appreciate their own strength. (*Hazleton Sentinel,* August 31, 1897b)

Much like over a century ago, the "English speakers" needed to develop a new inclusive narrative of the region and begin recognizing and celebrating the new immigrant cultures as part of the region's general narrative. Perhaps Hazleton and many areas in northeastern Pennsylvania should reflect on these historical narratives and be ready to work with new immigrants to develop a more inclusive community and heritage.

CHAPTER 7

Turning the Corner

Reviving the Memory of the Lattimer Massacre

Congressman Lou Barletta, the major proponent of the Illegal Immigration Relief Act, often points out that his family came from Italy several generations ago, and he and others call Hazleton a city of immigrants. So, while Hazleton is a city of immigrants, many of its inhabitants endorsed the city's anti-immigrant laws. The claim of having an immigrant background has allowed the xenophobic portion of the population to hide behind and justify the way they treat the new immigrants poorly, and get away with it.

At the beginning of our research in northeastern Pennsylvania, we thought that an archaeological survey of the Lattimer massacre site could be one way to foster a dialogue about labor, race, and immigration in the community. The project also could provide a vehicle to illuminate the wealth disparity that existed during the Gilded Age, as well as the increasing disproportion of the accumulation of wealth today (Roller 2015; Shackel and Roller 2012; Shackel 2016).

In the summer of 2010, Michael Roller and I set out to develop a plan for an archaeological survey of the Lattimer massacre site. Accounts of the massacre are highly contested, considering the sheriff and others filed conflicting accounts. Additionally, complete court transcripts from the trial are missing, though fragments of eyewitness accounts exist in newspapers. Palmer's autobiography (mentioned in chapter 3) has a portion of the court records transcribed, although they are mostly biased toward the defendants. The archaeological survey, with its material evidence, contributes an entirely new form of account to contest, affirm, or provide entirely new forms of evidence and bring attention again to the events surrounding the massacre. Therefore, our goals for this survey included finding the location of the initial engagement at the massacre site, and identifying a firing line for the posse and the coal and iron police (Roller 2015, 147; Shackel et al. 2011).

With the violent confrontation and the firing of munitions, it became clear to us that the place probably had similar archaeological signatures as other battlefield sites. For instance, using a metal detector, Doug Scott's survey of the Battle of Little Bighorn (Scott and Fox 1987; Scott et al. 1989) and the Sands Creek massacre (Scott 2003; Greene and Scott 2004) provide new insights into these confrontations. In addition, Larry Zimmerman worked with others to rewrite the history of the Outbreak of 1873 (McDonald et al. 1991). For a long time, these events relied heavily on military narratives that provided a skewed perspective associated with the American Indian wars west of the Mississippi. However, in some of these cases, the archaeological surveys provide new, alternative interpretations based on material evidence that coincide closely with American Indian oral histories.

Based on oral histories and newspaper accounts, the Lattimer massacre occurred at the western entrance to the town of Lattimer Mines. The sheriff and his posse, which were joined by the coal and iron police, stood guarding the mine and breaker located just north of the town entrance. Today, the land we surveyed is in private hands, owned by Pasco Schiavo, a prominent Hazleton attorney who has written about the region's history. He was raised in the Hazleton area and is a descendent of Italian immigrants who once lived in the Lattimer community. A visit to his law office allowed us the opportunity to discuss the project. Once we explained our intention to highlight the history of the massacre, Mr. Schiavo graciously allowed us to perform the survey on his property.

At first glance, the potential for finding archaeological material related to the massacre is daunting. Parts of the site were once covered by a culm bank, the waste material remaining after sorting coal from shale. A gravel road runs along the eastern portion of the site and leads to the most recent strip-mining operations north of Lattimer. Lining the west side of this gravel road is debris from Hazle Township street sweepers. At the massacre site, they dump their collected street waste (figure 7.1). The entire site is covered by secondary growth, which is predominantly acid-loving birch trees. The Lattimer massacre monument (or the Rock of Solidarity) is across the road at the town's entrance. It is located where the trolley tracks formerly lay. The raised bed for the tracks was removed after the 1950s, although the tracks' alignment is visible in the wooded area west of the site.

We had access to a map drawn by the coal company's engineer. Joe Michel, a former city engineer, purchased the Lattimer Coal Company records after

Figure 7.1. Street sweeper from Hazle Township dumps road debris at the massacre site. The remainder of the site is covered by secondary growth, which is predominantly acid-loving birch trees. (Photograph by Paul A. Shackel.)

the company closed. Michel grew up in a coal town known as Eckley, now a Pennsylvania historic site. Included in his archives are dozens of maps related to the region's coal industry. He graciously allowed us access to these records. The map drawn by the engineer immediately after the massacre shows the black gum, also known as the massacre tree, which marks the initial confrontation between the sheriff and the strikers.

Prior to our survey we were able to meet with members of the UMWA in their national office in Triangle, Virginia; however, we did not receive any response to our survey from local union representatives. We tried contacting them several times. Though commemorative religious services were held at the massacre site from the 1970s until about 2005, we thought the archaeological survey of the site might stir emotional pro-labor feelings in the region. Michael Shanks and Randall McGuire (1996) remind us that the act of archaeology is a form of commemoration, and when we do archaeology, we create a memory of the past that is rooted in our present-day concerns. Therefore, labor archaeology can be a way to remember and unveil a history that has been buried all too long. Faced with deteriorating conditions, labor organized to protect their rights as citizens, ensuring that they would receive a decent wage to support themselves and their families. An archaeology that emphasizes these issues can only help to remember and commemorate these histories.

Not having the expertise in metal-detecting surveys, Roller sent out a message on several listservs asking for advice and help for this potential survey. Within twenty minutes, he received a reply from Dan Sivilich, president of BRAVO (Battlefield Restoration and Archaeological Volunteer Organization) of Monmouth, New Jersey. BRAVO is a nonprofit organization that works to promote public interest in history and archaeology. Sivilich has family roots in the anthracite region close to Lattimer, and he was thrilled to offer his expertise to the project and bring his BRAVO team to aid in the research. Though there are newspaper accounts and fragments of the trial reporting, we saw this survey as a way to add a material component to the narrative.

Over two weekends in November and December 2010, Sivilich and his BRAVO team surveyed the area. Based upon historical accounts, photographs, and aerial photography, the team delineated a broad 6.5-acre survey area and identified the general location of the massacre tree. The tree served as a historical landmark until the late twentieth century, when a townsperson cut it down.

Initially, the crew followed systematic ten-foot transects in the project area. They then performed a semi-systematic survey of the area based on initial survey results. The survey on this tract produced forty-three artifacts, which included seven bullets, twenty-two cartridges, six copper jackets, a miner's tin cup with perforations from a shotgun blast, a cupric metal suspender clip, a silver-gilded serving spoon, and several objects of unidentified copper alloy hardware (Sivilich 2011). Artifacts were recorded with a total station laser transit, as well as a hand-held GPS unit to piece plot the location of each found object. All data were compiled and plotted using ArcGIS 10.2 software on geo-referenced aerial photographs (Sivilich 2011).

After collecting these artifacts, we consulted with Doug Scott for identification of the munitions. Law enforcement agencies have used different munitions identification techniques to help solve crimes. Identifying bullets and cartridge casings helps determine the type of weapon used at an incident. Specialists, like Doug Scott, can successfully link bullets and cartridge casings to weapons by identifying the firing pin, extractor marks, or the land and groove marks made by a rifle barrel during firing. Therefore, firearms identification (sometimes referred to as forensic ballistics) is similar to archaeologists examining wear pattern to identify use and functions of artifacts (Scott 2011).

Scott examined the munitions for tool marks and recorded the weights and diameters for each bullet. In many of these cases, unusual wear patterns were carefully documented to match munitions to a particular weapon or class of weapons (Scott 2011). It is difficult to identify munitions from the period of the massacre because of numerous developments in armament technology during this time, including the centerfire cartridge, smokeless powder, high-velocity rounds, and copper jacketing (Roller 2015, 151).

Most of the munitions uncovered date to after the Lattimer massacre and are probably related to recreational, subsistence-based, or crime-related gun activity. Scott identified sixteen shell casings; all date to after the massacre. Seven bullets were found and four could be related to the massacre (Scott 2011). The low number of bullets found associated with the massacre may be a result of souvenir collectors who came to the massacre site after the incident. The *Daily Standard* (September 20, 1897) noted that relic hunters carried away blood-stained pebbles and rocks as well as tree limbs pierced by bullets.

The type of bullets found near the massacre tree provide important evidence that there is a high likelihood they are associated with the event. These bullets include two .38 caliber long and short rounds. The manufacture of this bullet began in 1875 for the Colt revolver, although many other manufacturers subsequently used this caliber round for their revolvers. Markings on the bullets suggest they were probably fired from one or two different Smith & Wesson revolvers (Scott 2011, 4). A heavily impacted .22 round looks like it was smashed because it struck an object at high velocity. Its caliber was determined by its weight. The .22 was developed in the 1860s, and it remains the most popular small-bore round. These three rounds, the two .38 and the one .22, were found in a cluster at the edge of the road, adjacent to the plotted position of the massacre tree, as identified on the map drawn by the engineer the day after the event. There is a fourth bullet, a heavily patinated .32 caliber pistol round, which was found approximately 170 feet from the massacre tree. The .32 was introduced in 1875, and the groove marks suggest it was fired from a Smith & Wesson revolver. The heavy patination is similar to that of other period munitions identified (Roller 2015, 153–154; Scott 2011, 3) (figure 7.2). A drinking cup often used by miners was also found in the survey. It is riddled with shotgun pellets, and was probably dropped and lost in the mayhem during the shootings.

According to oral testimonies and artists' renderings of the event, there is some discrepancy about the placement of the posse during the confrontation with Sherriff Martin and the protesters. Sheriff Martin noted that he originally placed his men in a crescent shape formation that crossed the

Figure 7.2. Four bullets found as a result of the archaeological survey. These bullets date to the time of the Lattimer massacre: .32 caliber, .38 caliber, .22 caliber, .38 caliber. (Photograph by Paul A. Shackel.)

road. This formation would have confronted the unarmed marchers head-on with a force of Winchester rifles, pistols, and shotguns. Martin, however, rearranged the formation in a line that ran parallel to the fence line of the first house on the west end of town. The schoolteacher who witnessed the shooting from the Lattimer schoolhouse explained, "My attention was attracted to an unusual noise. I looked out and saw deputies coming on a car. The schoolchildren ran out. Then I saw the strikers coming down the road. The deputies got out of a car and formed a line. . . . It was partly across the road. Then they broke line and re-formed along the fence" (*Plain Speaker,* February 5, 1898). Andrew Yesmond, who was shot in the leg and spent twelve weeks in the hospital, testified that, "I fell on the same road that we came along. The firing came from the deputies along the fence" (*Daily Standard,* February 18, 1898). S. E. Jones, an alderman from West Hazleton, explained, "I saw the strikers on Broad Street on September 10 about noon. Saw them again at Lattimer, where I also saw the deputies' line along the road. I saw a scuffle between the strikers and the deputies. I was at a distance of about 75 to 100 yards" (*Daily Standard,* February 8, 1898). The line of men then ran into a field running westward and paralleled to the road (Martin, in Palmer 1913, 80; Novak 1977, 125). The formation created an enfilading line. Enfilade is a military term used to describe an attack or firing along the length of the target, rather than in a frontal assault fashion.

It is also known as flanking fire. Because of the curve of the road into the town, the line of deputies was about one hundred feet from the marchers on the right (westernmost portion of the line) and about fifteen feet away on the left (easternmost portion of the line (*Wilkes-Barre Times,* quoted in Novak 1977, 131; Roller 2015, 158). At the time of the confrontation, the four hundred striking workers were flanked by about 150 armed men to the north, and to their immediate south was a three-foot berm, the bed for the trolley tracks. The formation left the marchers in a tight place with little room to maneuver freely (figure 7.3).

The bullets found at the initial impact area of the massacre site include two .38 bullets from one or two Smith & Wesson revolver(s) as well as an impacted .22 bullet. The miner's drinking cup is also probably from the confrontation's initial impact. Though Martin explained that each of the

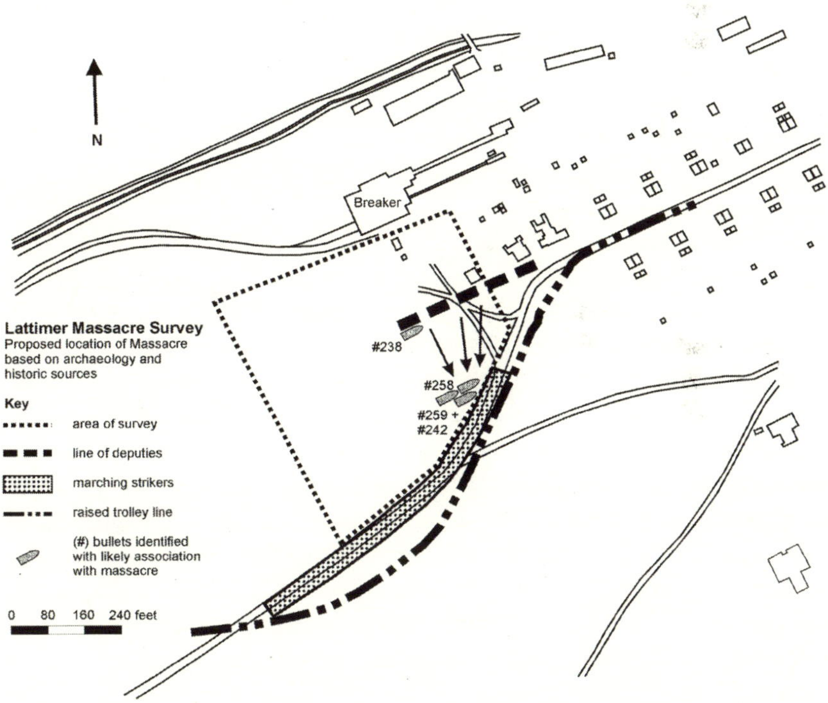

Figure 7.3. The formation by the sheriff's deputies created an enfilading line. The four hundred striking workers were flanked by about 150 armed men to the north, and to their immediate south was a three-foot berm, the bed for the trolley tracks. (Image by Michael Roller.)

deputies were provided Winchester rifles, he also noted that a few had buckshot guns (Martin in Palmer 1913, 83). We know that Martin was wielding a revolver, because he pointed it at one of the strikers. Perhaps a few of the men in the posse also had a revolver. Most likely the sixty to seventy men from the coal and iron police, company guards, and other law enforcement agents joined the posse to create a line of about 150 men (Roller 2015). The fourth bullet found at the right side of the deputy's line is a highly patinated .32 revolver one. Perhaps a frightened member of the posse fired it into the ground during the frenzy. Testimony at the trial by one of the witnesses provides some explanation for the presence of the small-caliber bullets. A fruit peddler, John Petruska, was in Hazleton, and he mentioned that on September 10, he saw deputy sheriffs purchasing guns in the hardware store. "I heard people inquire why the arms were being procured, and the men said that they were going to shoot" (*Daily Standard,* February 14, 1898). In a panic, members of the posse were arming themselves with any type of weapon they could. Perhaps only smaller pistols were left, and perhaps the ammunition from these pistols is what remained at the site.

The newspaper and other historic accounts indicate that after the initial firing, the deputy line continued to fire for between one and three minutes. It seems like the initial firing that impacted the front of the striking line came from those not armed with Winchester rifles, but rather by those with sidearms. The Winchester bullets are probably located outside the initial impact zone. Evidence of this firing may be found closer to the schoolhouse, or west of the initial impact area, where historical accounts note that the deputies chased and fired at protesters as they fled.

The archaeology survey at the Lattimer massacre site recalls the struggle between labor and capital and its material remains, highlighting the horrifying story of the racism that justified the merciless killings. This survey is one research program that is helping to reawaken the memory of the Lattimer massacre. It is illuminating the implications of race and racialization that enabled exploitation of the foreign-born working class, a phenomenon being operationalized today. These points were made in the local newspaper when we discussed the results of the survey. As a result, I received some negative responses from former Hazletonians via email. One person wrote, "Crime and murders/s are [now] common place. . . . My old high school has gangs!! This all happened when illegals from NEW YORK came to Pa. . . . Not to work, like the aliens of the past came, but to sponge off the social welfare programs and bring the big city crime of NYC. . . . Lou Barletta was trying to protect the interests of the citizens of Hazleton. . . . So please don't rush to judge the people of Hazleton" (email letter to the author, February 19, 2012).

The Future Meaning of Lattimer

In 1911, the *Reports of the Immigration Commission* (U.S. Senate 1911) sanctioned the supposed racial differences supported by the current science and entrenched in the American vernacular. It created clear boundaries between new and old immigrants. There are often references about the lack of mental and physical capacity among new immigrants (Handlin 1950, 82). Throughout this document's text are scattered phrases and sentences, reflections upon the lesser capacity of the new immigrants to be Americanized. The English and Irish came to the United States "imbued with sympathy for our ideals and our democratic institutions." The "Norse" make "ideal farmers and are often said to Americanize more rapidly than do the other peoples who have a new language to learn. . . ." For "the German is too well known in America to necessitate further discussion." By contrast, the Serbo-Croatians have "savage manners." Although the "Poles verge toward the 'northern' races of Europe," being lighter in color than the Russians, "they are more high strung, resembling the Hungarians" in that respect (U.S. Senate 1911, quoted in Roucek 1969, 44).

Just a few years earlier, in 1897, the miners of the Hazelton district were challenging the meaning of citizenship. Were these working immigrants equal citizens as they carried the symbol of American democracy and equality—the American flag? Or were they a subaltern group that could be exploited by the coal operators and the privileged citizens without contestation? Today, over a century later, there is a major influx of immigrant workers viewed by many Americans as not of equal status and a people not deserving of equal treatment as citizens, as the nativist movement becomes stronger in the United States. Though workers at many of the larger corporations, factories, and stores try to organize and bargain collectively, we are told that the market, rather than justice, should dictate the conditions of work. So, have things changed much since Lattimer?

Victor Greene, who addressed the Lattimer massacre in *The Slavic Community on Strike: Immigrant Labor in Pennsylvania Anthracite* (1968), laments how union supporters were racists and lobbied for immigration quotas, when they could have taken a very different tactic that would have helped the unions grow and prosper. He explains,

> Labor leaders then and perhaps now tend to overlook the fact that the work force is a part of America's pluralistic society. In the past, rather than trying to limit the numbers of component ethnic groups, for their own benefit unions should have accepted this country's heterogeneity. They should have conformed their tactics to the cultural habits of those that they wished to attract. Even today, by refusing to recognize the more self-conscious minorities,

> mass associations will stagnate, never to expand with the growing rank and file. (Greene 1968, 215)

I am hopeful that the tragic history associated with Lattimer, and the recalling of the unjust treatment of the new immigrant population at the time, can serve to remind us what it is like to be a newcomer to a foreign land, where people are suspicious of you because of your different customs and because you do not speak the native language. A descendant of Michael Cheslock framed the narrative of the strike. She said the miners were fighting for justice, a right they thought they had, and believed they were protected by carrying a symbol of justice, the American flag. "I mean, it must've been appalling what they worked under, you know, how they had to live and to survive. It was ongoing. The fact that they stood up for themselves, came over and stood up for themselves. They said 'we're not going to put up with that, we're going to strike.' I mean, they didn't realize they were risking their lives. They had the American flag" (interview, Mary Jo Cheslock Barrett, March 19, 2010).

I am also heartened that there is a growing awareness of the events surrounding Lattimer, and that many citizens in the coal region have used the incident to remind us about issues related to contemporary immigration. Newspaper editorials in Scranton, Hazleton, and Wilkes-Barre have responded to the anti-immigrant attitudes of government officials by calling upon the memory of Lattimer to teach about tolerance and justice. People write about their eastern European ancestors who met prejudice and resentment about a hundred years ago. One editorial writer stated, "Read what happened to your ancestors in the Lattimer Massacre. 'Those who do not learn from history are condemned to repeat it.' Accept all people of Hazleton as one people" (Klemlow 2009). Another letter in the editorial section of the *Standard Speaker* (Wittig 1993) connects the racism their grandparents faced at the turn of the century to the contemporary situation with new immigration. He recalls a recent incident where a Hispanic couple had their windshield smashed by a cinder block. "Haven't we learned anything since 1897? Wake up Hazleton. It's 1993. Learn from the past. Remember how you and your neighbor's grandparents felt and stop treating immigrants to our town as criminals just because their skin is a different shade. Let's correct the mistakes of the past instead of repeating them" (Wittig 1993).

In an oral history, John Probert, a multiple-generation Hazletonian, believes the incident at Lattimer has major national significance related to labor and class issues. Yet part of why he thinks it is so significant is that the principles involved in the massacre are resurfacing.

> You don't have to look too far back, go back a year, not even a year . . . [and] look at the financial meltdown. Look at that! That's exploitation! Same kind

> of a thing, same concept. Those that have are exploiting those that don't . . . it's all through history you see the same thing. As soon as one group of people get the ability to exploit another for their own benefit, they do it.

He also notes:

> . . . And it doesn't matter which age you're in, they just do it by different means. Slavery, of course, being one egregious example of it, but there are many others! Now I've often said, the closest the white man ever came to slavery in this country was right here in this region. Yes, they weren't slaves; they were paid, but they were essentially slaves. They had very little options. They had to work for that coal company. (Interview, John Probert, March 19, 2010)

Probert understands the massacre to be, at least in part, an issue of economic and social exploitation—something he feels can be seen throughout history, including today.

From the early 1970s until about 2005, St. Mary's Church in Lattimer supported a Mass at the monument site on the massacre's anniversary. The Mass was complemented with a breakfast and small parade. Around 2005 the commemorative Mass ceased, and the parish merged with several others in 2009. The community showed little interest in reviving the commemorative Mass. However, 2013 brought an interfaith service at the site during the anniversary (Whalen 2010; Ragan 2013).

Father Generose has been a major player in recent years commemorating the Lattimer massacre. He justifies the church support of commemorative activities with the following sentiment: "The Lattimer massacre and the events that led up to it are just part of that great labor movement during the industrial revolution where people are trying to protect the dignity of their own lives and the church is strongly behind that. . . ." He explained that the ecumenical event on the anniversary of the massacre is important in a local as well as global context. "We'll have the Lutherans there, we'll have maybe the Episcopalians there to remember, but it is in the general context of peace and justice that we do it and how it's so urgently needed in our world today" (interview, August 20, 2014).

"Teach PA History" has developed a teaching module related to the Lattimer massacre, along with other histories related to labor and the anthracite region (ExplorePAhistory.com).

The Growing Need for NGOs

With the city's recurring financial challenges, and now having to pay court costs for pursuing the Illegal Immigration Relief Act, the City of Hazleton lacks adequate resources to support all the needs of its citizens. With the development of the tax-friendly trade zones and the growing need for low-wage workers, the city of Hazleton now has a Latino population approaching 50 percent. Many pursue the available low-paying, unskilled, blue-collar jobs, exactly what the community leaders have created with these low-tax commercial zones. A large number of newcomers work in the fulfillment centers of big-box stores, in light manufacturing, or in meatpacking. Many find work through temporary employment services, and the workers receive little or no benefits, such as health insurance and unemployment compensation. These individuals earn relatively low wages, which explains why about 67 percent of students in the Hazleton Area School District come from families with an annual income below the poverty line.

There is an urgent need to assist and provide services to the city's underserved population. Small, nonprofit organizations have developed to fill this urgent need, some achieving great success. There is Hazleton POWER (Professionals Organized & Working to Enrich the Region). Its mission is to connect local professional organizations with the community through social-wellness campaigns and volunteerism.

Recently, the Hazleton Integration Project (HIP) developed as a response to the lack of city support for children of all backgrounds in Hazleton's underserved areas. The project developed with the leadership of Joe Maddon, manager of baseball's Chicago Cubs, and his family members and close friends. Maddon grew up in Hazleton and became distraught after seeing the racial tension in his hometown making national news (Hazleton Integration Project 2016).

The HIP mission statement notes the importance of different cultural groups spending time together to truly get to know each other while understanding and appreciating cultural similarities and differences (Hazleton Integration Project 2016). Maddon spent his adult life working with, coaching, and managing Latino ballplayers, and in media interviews, he explained that they shared the same goals and values as many Americans. What motivated Maddon to get involved was his experience socializing with Hispanic friends of the family while visiting Hazleton. He recognized the close family bonds, which reminded him of his experience growing up in Hazleton. Maddon thought it important to create a community whereby newcomers enjoyed opportunities to socialize with the traditional community. Therefore, Maddon and others founded the Hazleton Integration Project as a way to promote positive relationships between different groups through community activities (Hine 2015).

The local community and politicians did not immediately embrace HIP. For instance, Jim Grohol, owner of Jimmy's Quick Lunch in downtown

Hazleton, reported that some of his customers were not fond of what Maddon is trying to do for the underserved population. "The talk over the counter—that's exactly what they say," Grohol said. "They say, 'Oh he doesn't live here. He doesn't know what it's like to live here. He needs to find out'" (Hine 2015). Some cynics of the HIP project have labeled Maddon an enabler.

Maddon responded, "There has been a lot of crime going on in Hazleton forever and there are people who have been born and bred here that have been creating a lot of problems. . . . Just don't blame it on this group coming in" (Hine 2015). The retired Hazleton police chief, Frank DeAndrea, confirmed that most of Hazleton's older, white population choose to forget that Hazleton has a history of crime. In the 1970s and 1980s, it was known as "Mob City" (Hillyard 2017). Mafia members, escaping the watchful eye of the FBI in New York City, came to Hazleton. Stories and a few newspaper accounts tell of mob-related incidents during this era.

HIP purchased the former Catholic school Most Precious Blood and named the project and building Hazleton One. The organization took twenty-five classrooms and transformed them into an active educational space. One classroom is dedicated to teaching English as a second language (Hine 2015). Many of the programs are geared toward children. There is now a newly refurbished basketball court and a playground. Hazleton One offers music classes, and there is an industrial-sized kitchen in the basement, which helps provide meals to families in need. Reaching out to the underserved children is important to Maddon. "The kids have to learn how to not like somebody else," Maddon said. "They normally have to be taught by their parents. All the grown-ups, any of this prejudice, any of this crap going on is learned behavior" (Hine 2015). In a city with few resources to help the underserved population, the Hazleton Integration Project came forward to provide needed resources for community youth.

Local businesses owned by the traditional white population are doing better financially because of the connection some have made with the Hispanic community. "Money spends the same regardless of who's spending it," Maddon said of the local economy. "So I think a lot of the Anglo owners in the area are totally benefiting from the Hispanic population being here" (Kester 2016). Because of its initial success, several universities are studying the center, and may use it as a model for other distressed cities (Kester 2016).

At the Hazleton One Community Center I enlisted high school students to participate in our summer archeological field school. During the first day of summer recess for the Hazleton Area School District, I was heartened to see dozens of cars lining up and dropping off their children, encouraging them to take advantage of various educational and recreation programs. The University of Maryland sponsored an archaeology program in which students participated in our excavations. During the summers of 2016 and 2017, we were able to recruit six students from the Hazleton Area School District to participate in our archeological excavations. Three of the students were of Hispanic background and came with the encouragement of the Hazleton One board members, Bob Curry and Elaine Maddon Curry. On several occasions, we went to the community center on Saturday and presented a program about archaeology, followed by an overview of some of the artifacts we found in a previous excavation season. While bringing some of the items we found in an excavation, like pipe stems, fragments of plates, and marbles, we also brought along a few Catholic artifacts the Latino students could connect with, like part of a rosary, a portion of a porcelain crucifix, and a medal with an inscribed figure of Papa Pio IX, which is the Italian for Pope Pius IX (who reigned as pope from 1846 to 1878, the longest-serving elected pope). Many of these objects show a connection between the values of the historic culture and values of the new immigrants.

Our archaeology project continues to focus on immigration issues. Our message to the public was that our project explored the lives of the earliest European immigrants to the region. Though the new immigrants of the late-nineteenth century faced many hardships—like language barriers, different customs, and filling the lowest-paying jobs—these are the same challenges the new Latino immigrant faces today in northeastern Pennsylvania.

Therefore, one of our goals for the Anthracite Heritage Project is to develop universal meanings on the heritage landscape, which will allow us to connect past narratives of immigration and racialization to current social and economic conditions in the region. The emphasis on the heritage of immigration can potentially bridge a common experience between the established European-descendant populations and the new Latino immigrant. So, while we are not proposing to reverse the narrative, we are trying to build bridges between disparate communities, with the ultimate goal of using heritage to create a bridging social capital in a racially tense and poverty-stricken region. Bridging social capital refers to the networking between socially heterogeneous groups, which tends to create many benefits for societies, governments, individuals, and communities (Putnam 2000; Little and Shackel 2014).

The recruitment of high school students became a critical part of our community outreach to help bridge the meaning of the past to the present between these different communities. The Hazleton-area school system consistently ranks near the bottom in the Commonwealth of Pennsylvania, regularly placing in the low 400s out of 498 school districts. High school scores in math, reading, and writing show a steady decline over the past five years, but science scores are holding steady (*Pittsburgh Business Times* 2012).

Our archaeology program consisted of a six-week college-level field school. During the fourth week, high school students joined the program and were mentored by college students. Our Hazleton One students were first-generation Americans with connections to the Dominican Republic. While working side by side with their college student mentors, they also worked with other white high school students, who were of eastern European descent. By providing this type of working environment, students socialized with others outside their comfort zone and learned a bit of archaeology. They quickly learned to excavate, take notes, and curate artifacts. While we often discussed issues related to immigration, the partnering of college students with high school students led to many other discussions, mostly surrounding issues related to higher education. There were discussions about GPAs and taking seriously the SATs, which would help these students get into good schools and increase their chances of earning scholarships. In many ways, this mentorship program—high school students working with college students—may have served as a better experience and a freer exchange of information than any guidance counselor in high school.

While Joe Maddon brings his celebrity power to Hazleton for HIP fundraisers, the organization also relies on corporate donations to help provide programs and develop new resources. For many NGOs that rely on corporate sponsorship to help underserved populations, the act of corporate giving is necessitated by the low wages these local companies provide their employees. Corporations can pay low, subpoverty wages, and yet when they donate to the NGO to help underserved children, they are seen as partners, contributing to the needs of the community. Though hers is not a true form of corporate giving, I am reminded of the local stories about Sophia Coxe, the wife of one of the major coal operators in northeastern Pennsylvania. She was often referred to as the Angel of the Anthracite because of her philanthropy, providing the miners' children with clothing, toys, and perhaps an orange during Christmastime. In one instance, she provided a maimed worker with a sewing machine so he could pick up a new trade. One local told me she would leave uneaten items in her picnic basket after an outing with family or friends so that locals could supplement their meager diets. These acts created a mythology of her philanthropy. Would there have been a need for her giving if her husband's corporation paid a living wage? In reference to the Lattimer strike, the *Daily Standard* (September 1, 1897) wrote regarding the coal operators' attempt at philanthropy: "There is not much philanthropy in paying the ignorant foreigners 90 and 95 cents a day and then rob them of even this small pittance through the company store and company butcher. True philanthropy only exists in advancing the conditions of the poor toiler instead of degrading it." In a contemporary setting, Longazel (2016, 82) provides a perspective about corporate giving to HIP. The gesture "reinforces neoliberal ideology by encouraging the continued existence of an organization that does not threaten the prevailing economic order and that has filled the socially devastating void that is the result of neoliberal cuts to state-funded social services" (Longazel 2016, 82).

Changing Tide

Engaging communities has become an important part of many research designs, with the goal of developing a more inclusive heritage. Inclusiveness is about developing a more complex narrative that includes many voices and a variety of interest groups. One strategy in developing a more inclusive narrative is to develop universal meaning(s) whereby various groups of different backgrounds can identify with the meaning of a particular heritage. Therefore, an important decision in northeastern Pennsylvania is whether to emphasize the local and a community's unique historical characteristics, or to highlight a community's heritage focusing on universal values connecting disparate groups. Developing universal meaning for a place allows us to not only connect many different interest groups, it also allows us to make links between the past and the present, which can facilitate an exploration of both historic and contemporary concerns related to social justice (AAM 2002). Though the outcomes of exploring universals may prove productive, these results are neither automatic nor guaranteed.

Culturally specific museums that attempt to break down traditional boundaries and explore similarities and differences between groups have provided a new way of making important links. One example is the National Museum of Mexican Art in Chicago. Curators worked with members of the local Latino and African American communities and developed an exhibition titled, "The African Presence in Mexico: From Yanga to the Present." This work heightened the community's awareness about the complex history of race and ethnicity in Mexico. In Seattle, the Wing Luke Asian Museum is a cultural center that serves the Asian Pacific American community. Its goal is to promote cross-cultural understanding among the many different groups and nationalities categorized as "Asian," promoting the idea that there are many different groups in this category, each with their own distinctive cultural norms and values (Farrell and Medvedeva 2010, 20).

The Levine Museum of the New South in Charlotte responded to Robert Putnam's (2000) report on social capital, which identified Charlotte as having a low level of intergroup trust. The museum developed exhibitions specifically around issues of race, racism, and trust. In 2004, its first effort to address this issue of trust was titled "Courage." The exhibition addressed the history of school desegregation. Working with a community group, the museum brought in professionals from across the city to develop focused discussions on this topic. "The hallmarks of the project are engaging and provocative questions that get people talking about tough issues: Who judges you without knowing you? Who do you judge? What parts of your cultural heritage have you kept? Let go of? What cultural aspects of the South most surprised you?" (Farrell and Medvedeva 2010, 20). The museum used a historical event to develop a deeper understanding of some of the challenges facing this community. Because of the exhibition's success as a catalyst for community dialogue, the commitment to community is now an important mission of the museum. The Levine Museum of the New South has taken a leadership role in their region for developing civic programs that address demographic changes (Farrell and Medvedeva 2010, 20).

A prominent example of developing a universal meaning of a historic phenomenon and linking the historic to the present is the Tenement Museum in New York City. It is a place that helps visitors recognize the experience of the historical as well as the new immigrant, and tries to guide visitors to identify with historical and contemporary experience. The immigrant experience is one condition that unites people across time and culture. Those involved in the museum believe historical discussions of immigrants' lives can create a sense of empathy and tolerance for the new immigrant (Russell-Ciardi 2008, 42).

The Hazleton community and many others in northeastern Pennsylvania have a long and proud history of mining. The region's history relates to immigration and climbing the ladder to success. Each set of newcomers faced xenophobic resistance because they spoke a different language, observed different customs, and/or dressed differently. They were also challenged with harsh working conditions, infrequent work, and substandard living conditions. The new Latino immigrants are facing the same types of discrimination the eastern and southern Europeans received several generations earlier. They have the lowest-paying jobs and the worst access to health care, and are demonized as criminals and accused of not wanting to assimilate into American culture.

Today, Broad Street, Hazleton's main avenue, contains a few remaining traditional businesses established by the white population several decades ago, like Jimmy's Quick Lunch and Hazel Drugs. Interspersed are a growing number of enterprises—restaurants, clothing stores, and botanicas—that cater to the new Latino population. A Latino family now operates the Italian restaurant. The *New York Times* reported that "Hazleton's old shopping streets, nearly abandoned in the 1990s, are now lined with Dominican bakeries, barbershops, travel agencies and Mexican restaurants. . . . The city has two Spanish radio stations and a television station that broadcasts six hours of local programming a day" (Appelbaum 2016). The one hundred or so new businesses have played a significant role in helping revitalize the community's economy. There is now a bit of hustle and bustle in the downtown area not seen for several decades. "'With the influx and change in population, they come here and open small businesses—it's the same thing that happened 100 years ago,' said Mayor Jeff Cusat, standing on a downtown Hazleton street corner as he noted the changes to the city's storefronts" (Hillyard 2017). The city's downtown is vibrant, especially when compared to the many other faltering communities in the anthracite region.

Recent newspaper articles in the *Philadelphia Inquirer* and coverage by NPR no longer mention the Illegal Immigration Relief Act, but rather describe the revitalization of the downtown area with new merchants catering to the Latino population, as well as programs like HIP (Klibanoff 2015; Matza 2016). Dr. Schleicher, a lifelong Republican who practices in downtown Hazleton, said, "We're seeing a total revitalization despite the government trying to keep the immigrants out. It would have been a ghost town of older white people" (Appelbaum 2016).

Organizations like the Downtown Hazleton Alliance for Progress (DHAP) are working to fight against blight and crime. They work to support cultural institutions as well as finding new uses for historic buildings. The group is trying to create a perception of a healthy downtown area with events like a farmers' market, first Fridays, and movie at the Markle. The streets and sidewalks have been replaced along Broad Street, and three larger building are being renovated for office space (Blanco 2016). For the first time in decades, a major employer is locating its headquarters on Broad Street. DBi Services purchased and restored the Traders Bank and Hazleton National Bank buildings, and will bring employees to the downtown corridor. The company designs IT infrastructures for companies.

With the influx of a new population, there is an economic and infrastructure revitalization underway. Some residents have decided to welcome the newcomers, like Bob Curry and Elaine Maddon Curry, who are spearheading the efforts of Hazleton One. They remember the stories about the struggling new immigrant, and in their view Lattimer is an important example about what people endured in order to protest for decent wages and working conditions. Others have grudgingly accepted the new immigrant as they feel the pressure of rising rents and a general suppression of wages. "'The majority of the Latinos here are hard workers and they're doing a lot of good for the city,' said Mary Kirkpatrick, a local housekeeper, as she sipped on a watery Bud Light and puffed away on a USA Gold cigarette at Henry's Bar on Broad Street. 'Unfortunately [immigration] has raised our rents and wages have gone down, so that's why Latinos are getting blamed for the city's problems,' she said. 'I've never worked for minimum wage in my life until now, but I can't blame immigrants as I would do the same things if I had to'" (O'Reilly 2013).

The 2016 presidential election, however, has changed the general tone of the community, and the anti-immigration rhetoric has once again embroiled the city. The Hispanic population continues to grow in Hazleton, leaving many of the older, white community members feeling threatened as their community changes and economic prosperity eludes them. "I don't care for this town no more because of the Hispanics," said Lewis Beishline, seventy, as he sat drinking on a Friday at Cusat's Cafe, a bar owned by the mayor of Hazleton, who lives upstairs. Mr. Beishline, a retired welder, said he moved from Hazleton to a nearby town last year because he no longer felt safe. He planned to vote for Mr. Trump, he said, "because of the immigration" (Appelbaum 2016). In another case, a woman who earns $11.50 as a home health aide is bitter about her situation, explaining that the surplus of immigrant labor keeps the general wages low. "No one talks about white Americans and what we really need," she said (Appelbaum 2016).

After the election, Latino residents in Hazleton who are here legally, as well as those who are U.S. citizens, explain that the rhetoric intended for undocumented immigrants is affecting the entire Latino community. Niurka De La Rosa, owner of a daycare center in downtown Hazleton, said that "If [Trump] continues the way he's been doing so far, it's going to be a big impact on those kids—even though they're second- or third-generation because everyone is going to label them: 'Go back to where you came from'" (Hillyard 2017). While Tea Party flags that state "Don't Tread on Me" fly throughout the city of Hazleton, on occasion I have also seen the "White Pride, Worldwide" flag flying in front of a residence in Lattimer (figure 7.4).

Hazleton has been a regional focal point for the anti-immigration movement since the mid-2000s. People have claimed the new immigrants (many

Figure 7.4. Tea Party flags with the "Don't Tread on Me" symbols are found throughout the Hazleton area. This "White Pride Worldwide" flag occasionally flew in the front yard of a residence in Lattimer. (Photograph by Paul A. Shackel.)

of them incorrectly labeled as Mexican) have brought drugs, increased crime, and taken jobs away from the white population. Ironically, Hazleton now houses several international Mexican-owned businesses, such as Bimbo Bakeries USA (maker of Thomas' English muffins), Mission Foods, and Wise. These companies employ a significant number of local residents (Blanco 2016). Francisco Torres-Aranda, a member of the Governor's Advisory Commission on Latino Affairs, lives close to Hazleton. He said, "'People were blinded by their fear of change, unable to see the benefits that immigration is bringing to Hazleton.' Mr. Torres-Aranda, whose father was Mexican, runs a company that makes caps for old wells. He employs 30 people in the summer, only a few of them Hispanic" (Appelbaum 2016).

Dr. Agapito Lopez, a retired ophthalmologist and an activist for the Latino community, knows what is at stake with the most recent election and the reigniting of anti-immigration sentiment in the city. "Hazleton is a mirror for the whole nation to look into," he said, "This is where America could wind up" (Alvaro Guzman 2016).

There are many parallels between racial attitudes of the past and the present. Working and labor conditions in the anthracite mines in northeastern

Pennsylvania are a pale memory in the official history of the United States. The ethnic tensions between the established "English-speaking" population (mostly of German, Irish, Welsh, and English descent) and the "newcomers" from southern and eastern Europe in the late-nineteenth century are viewed by many as a distant memory with few ramifications for those living in the community today. The silence surrounding these historic tensions related to class, poverty, and racism is quite deafening, especially considering the region is facing many of these same issues today with a new immigrant population. However, there is a story to tell and lessons to be learned related to one of the worst acts of violence in American labor history—the Lattimer massacre. The story has implications for how we remember the past and which past we choose to forget. What we remember, and the meaning of the past that we embrace, have ramifications for our everyday lives.

REFERENCES

Alcock, Susan E. 2002. *Archaeologies of the Greek Past: Landscapes, Monuments, and Memories*. Cambridge, England: Cambridge University Press.

Allan, Theodore W. 1994. *The Invention of the White Race,* vol. 1. London: Verso.

Alvaro Guzman, Bastida. 2016. "In Hazleton, PA, Latinos Brace for Elections Outcome." *AlJazeera*. http://www.aljazeera.com/indepth/features/2016/11/hazleton-pa-latinos-brace-elections-outcome-161107100957410.html.

American Association of Museums (AAM). 2002. *Mastering Civic Engagement: A Challenge to Museums*. Washington, D.C.: American Association of Museums.

American Civil Liberties Union. 2015. *Lozano v. Hazleton*. February 5, 2015 update. American Civil Liberties Union. https://www.aclu.org/cases/lozano-v-hazleton.

Anderson, Benedict. 1991. *Imagined Communities: Reflections on the Origins and Spread of Nationalism*. London: Verso Press.

Anthracite Coal Commission. 1903. *Report to the President on the Anthracite Coal Strike of May–October 1902*. Anthracite Coal Commission. Washington, D.C.: Government Printing Office.

Aurand, Harold W. 1971. *From the Molly Maguires to the United Mine Workers: The Social Ecology of an Industrial Union, 1869–1897*. Philadelphia: Temple University Press.

———. 2002. "The Lattimer Massacre: Who Owns History? An Introduction." *Pennsylvania History: A Journal of Mid-Atlantic Studies, Special Issue: The Lattimer Massacre (1897)* 69 (1): 5–10.

———, editor. 2002. *Pennsylvania History: A Journal of Mid-Atlantic Studies, Special Issue: The Lattimer Massacre (1897)* 69 (1).

Bahadur, Gaiuntra. 2006. "Hazleton Gets a Jolt it Didn't Want." *Philadelphia Inquirer*. www.smalltowndefenders.com/node/40.

Baker, Lee D., and Thomas C. Patterson. 1994. "Race, Racism, and the History of U.S. Anthropology." *Transforming Anthropology* 5 (1&2): 1–7.

Baratta, Lisa. 2007. "Judge Strikes Down Hazleton's Illegal Immigrant Law." *Pennlive.com*. http//blog.pennlive.com/midstatebuz/print.html.

Beik, Mildred Allen. 2002. "The Significance of the Lattimer Massacre: Who Owns Its History?" *Pennsylvania History: A Journal of Mid-Atlantic Studies, Special Issue: The Lattimer Massacre (1897)* 69 (1): 58–70.

Benton, Tim, ed. 2010. *Understanding Heritage and Memory*. New York: Manchester University Press.

Bernardo, Richie. 2016. "2016's Most and Least Educated Cities." *Wallethub.com*. https://wallethub.com/edu/most-and-least-educated-cities/6656/#methodology.

Blanco, Octavio. 2016. "How Latinos are Saving this Former Pennsylvania Mining Town." *CNNMoney*, September 2. http://money.cnn.com/2016/09/02/news/economy/hazleton-pa-latino-immigrant/.

Blatz, Perry K. 1994. *Democratic Miners: Work and Labor Relations in the Anthracite Coal Industry, 1875–1925*. Albany: State University of New York Press.

———. 2002. "Reflections on Lattimer: A Complex and Significant Event." *Pennsylvania History: A Journal of Mid-Atlantic Studies, Special Issue: The Lattimer Massacre (1897)* 69 (1): 42–51.

Blight, David W. 2001. *Race and Reunion: The Civil War in American Memory*. Cambridge, Mass.: Belknap Press of Harvard University.

Bluestone, Barry, and Bennett Harrison. 1982. *The Deindustrialization of America: Plant Closings, Community Abandonment, and the Dismantling of Basic Industry*. New York: Basic Books.

Bodkin, Karen. 2000. *How Jews Became White Folks and What that Says About Race in America*. Brunswick, N.J.: Rutgers University Press.

Bodnar, John. 1992. *Remaking America: Public Memory, Commemoration, and Patriotism in the Twentieth Century*. Princeton, N.J.: Princeton University Press.

Boric, Dusan, ed. 2010. *Archaeology and Memory*. Oxbow, England: Oxbow Books.

Breene, T. H., and Stephen Innes. 1980. *"Myne Owne Ground": Race and Freedom on Virginia's Eastern Shore, 1640–1676*. New York: Oxford University Press.

Brennan, John C. 1972. " '. . . They Are in Peace, Alleluia.' Eulogy by John C. Brennan, Secretary-Treasurer of Local 1361, Retail Clerks International Association, AFL-CIO." *United Mine Workers Journal*. Xerox copy, on file, Washington, D.C.: Labor Heritage Foundation.

———. 1978. "News Release—Mass at Labor Memorial. August 20, 1978." On file, Washington, D.C.: Labor Heritage Foundation, GMMA 98-002 34/19.

Brinton, Daniel G. 1890a. *Essays of an Americanist*. Philadelphia: Porter and Coates.

———. 1890b. *Races and Peoples: Lectures on the Science of Ethnography*. New York: N. D. C. Hodges.

Brooks, John Graham. 1898. Notes. *The Yale Review: A Quarterly Journal for the Scientific Discussion of Economic, Political and Social Questions*, vol. 6, pp. 306–11. New Haven, Conn.: Tuttle, Morehouse & Taylor Press.

Brown, Anna, and Mark Hugo Lopez. 2013. "Mapping the Latino Population, By State, County and City." Pew Research Center's Hispanic Trends Project. Washington, D.C.: Pew Research Center. www.pewhispanic.org/files/2013/08/latino_populations_in_the_states_counties_and_cities_FINAL.pdf.

Brunner, Edward. 1986. "Ethnography as Narrative." In *The Anthropological Experience*. Victor W. Turner and Edward M. Brunner, editors, pp. 139–55. Urbana: University of Illinois Press.

Campomizzi-Clews, Kathryn. 1997. "Lattimer Massacre 100th Anniversary Nears: Scholars to Assess Key Labor History Event at Sept. 12–13 Eckley Gathering." *Republic Herald* (Pottsville, Pa.) September. Alice Hoffman Collection. State College: Penn State University Library.

CBS Evening News. 2007. "Immigration Showdown in Hazleton. *CBS Evening News*. https://www.cbsnews.com/news/immigration-showdown-in-hazleton/.

"Cesar Chavez at Lattimer, 1972." Lattimer Massacre Project. https://lattimermassacre.wordpress.com/2014/10/22/cesar-chavez-at-lattimer-1972/.

Chandler, Alfred D. 1972. "Anthracite Coal and the Beginning of the Industrial Revolution in the United States." *Business History Review* 46 (2): 141–81.

Chavez, Cesar E. 1972. Letter, "Dear Brothers and Sisters." Lattimer file, Hazelton, Pa.: *Standard-Speaker*.

Chavez, Leo R. 2008. *The Latino Threat: Constructing Immigrants, Citizens, and the Nation*. Stanford, Calif.: Stanford University Press.

Chishti, Muzaffar, and Claire Bergeron. 2014. "Hazleton Immigration Ordinance that Began with a Bang Goes Out with a Whimper." Migration Policy Institute. March 28. http://www.migrationpolicy.org/article/hazleton-immigration-ordinance-began-bang-goes-out-whimper.

Cornell, Robert J. 1957. *The Anthracite Coal Strike of 1902*. Washington, D.C.: Catholic University Press.

Coxe, Robert. 1913. "In the Court of Oyer and Terminer, Luzerne County, PA., Commonwealth of Pennsylvania, vs. James Martin and Others. Trial of James Martin, Sheriff, and Others for the Murder of Mike Ceslak (1898)." In Henry W. Palmer, *Fifty Years at the Bar and in Politics,* pp. 209–65. Williamsport, Pa.: Snyder & Bischof.

Culen, Konstantin. 1977. "The Lattimer Massacre." *Slovakia* 27 (50): 44–61.

Curilla, George J., Jr. 1972. "Lattimer Massacre Memorial: Labor Groups Dedicate Monument." *United Mine Workers Journal*. News clipping, Washington, D.C.: Labor Heritage Foundation.

D'Agostino, Peter. 2002. "Craniums, Criminals, and the 'Cursed Race': Italian Anthropology in American Racial Thought, 1861–1924." *Comparative Studies in Society and History* 44 (2): 319–43. DOI: https://doi.org/10.1017/S0010417502000154.

Deasy, George F., and Phyllis R. Griess. 1961. *Tourism for the Anthracite Region: An Alternative for Unemployment. Mineral Industries* 30 (7):1–8.

DeFazio, Salvador. 2007. "Hazleton: 1857–2007." Program for Hazleton 150th Anniversary.

Dempsey, Bobbi. 1997. "Lattimer Massacre Observed: Marchers Note 100th Anniversary of Labor Dispute that Led to Deaths of at Least 19 Coal Miners." *Times Leader,* September 13. On file, Wilkes-Barre, Pa.: Osterhout Free Library.

DiAngelo, Robin. 2011. "White Fragility." *International Journal of Critical Pedagogy* 3 (3): 54–70.

Digger. 2007. "Hazleton Heads to Court Against ACLU, U.S. Chamber, Illegal Aliens, U.S. Bishops [day 1 Recap]." Diggers Realm Blog, March 14, 2007. http//www.diggersrealm.com/mt/archives/002106.html.

Dobbs, Lou. 2007. "Hazleton's Illegal Immigration Relief Act Has Been Struck Down—Appeal Likely." Posted, The Official Website of Representative David Duke, PhD—http.davidduke.com. http://www.davidduke.com/general/hazeltons-illegal-immigration-relief-act-struck-down-appeal-likely_2508.html.

Dublin, Thomas. 1998. *When the Mines Closed: Stories of Struggles in Hard Times*. Ithaca, N.Y.: Cornell University Press.

Dublin, Thomas, and Walter Licht. 2005. *The Face of Decline: The Pennsylvania Anthracite Region in the Twentieth Century*. Ithaca, N.Y.: Cornell University Press.

Dubofsky, Melvyn. 2002. "The Lattimer Massacre and the Meaning of Citizenship." *Pennsylvania History: A Journal of Mid-Atlantic Studies, Special Issue: The Lattimer Massacre (1897)* 69 (1): 52–57.

Duke, David. 2007. The Official Website of Representative David Duke, PhD. http://www.davidduke.com/date/2007/07.

Dulles, Foster Rhea, and Melvyn Dubofsky. 1984. *Labor in America, A History*. 4th edition. Arlington Heights, Ill.: Harlan Davidson.

Englund, Jason. 2007. "Notes: Small Town Defenders or Constitutional Foes: Does the Hazleton, PA Anti-Illegal-Immigration Ordinance Encroach on Federal Power?" *Boston University Law Review* 87 (4): 883–909.

Falchek, David. 2015. "Report: NEPA Region Lags in Advanced-Skilled Jobs." *CitizensVoice.com,* February 3. http://citizensvoice.com/news/report-nepa-region-lags-in-advanced-skilled-jobs-1.1826843.

Falk, Candace, ed. 2003. *Emma Goldman: A Documentary History of the American Years: Made in America, 1890–1901,* vol. 1. Berkeley: University of California Press.

Farrell, Betty, and Maria Medvedeva. 2010. *Demographic Transformation and the Future of Museums*. Washington, D.C.: The AAM Press, American Association of Museums. http://www.aam-us.org/docs/center-for-the-future-of-museums/demotransaam2010.pdf.

Feagin, J. R. 2006. *Systematic Racism: A Theory of Oppression*. New York: Routledge.

Finan, Nell. 2013. Blog post. Lattimer Massacre Project, Employee Record Cards, Part II. September 3. https://lattimermassacre.wordpress.com/2011/12/21/employee-record-cards-part-ii/.

Fine, Michelle. 2004. "Witnessing Whiteness / Gathering Intelligence." In *Off White: Readings on Race, Power, and Society*. M. Fine, L. Weiss, L. Powell Pruitt, and A. Burns, editors, pp. 245–56. New York: Routledge.

Fink, Leon. 2015. *The Long Gilded Age: American Capitalism and the Lessons of a New World Order*. Philadelphia: University of Pennsylvania Press.

Foner, Philip S. 1964. *History of the Labor Movement in the United States*. Vol 3. New York: International Publishers.

Foulke, C. Pardee, and William G. Foulke. 1979. *Calvin Pardee, 1841–1923: His Family and His Enterprises*. Philadelphia: Drake Press.

Frisch, Michael. 1990. *A Shared Authority: Essays on the Craft and Meaning of Oral and Public History*. Albany: State University of New York Press.

Gallup-Healthways Well-Being Index. 2014. *State of American Well-being: 2013 State, Community, and Congressional District Analysis. Gallup-Healthways Well-Being Index*. http://www.well-beingindex.com/.

Gass, Jonathan. 2007. "Immigrant Law Ruling Divides Hazleton." *Pennlive.com*. http//blog.pennlive.com/patriotnews/2007/07/immigrants_crackdown.

Glaeser, Edward L., Joshua D. Gottlieb, and Oren Ziv. 2014. "Unhappy Cities." *NBER Working Paper Series*. Cambridge, Mass.: National Bureau of Economic Research. http://www.nber.org/papers/w20291.

Glassberg, David. 1990. *American Historical Pageantry: The Uses of Tradition in the Early Twentieth Century*. Chapel Hill: University of North Carolina Press.

———. 1996. "Public History and the Study of Memory." *The Public Historian* 18 (2): 7–23.

Gompers, Samuel. 1898. "Samuel Gompers on the Lattimer Slaughter." *New York World*. February 20.

Greene, Jerome A., and Douglas D. Scott. 2004. *Finding Sands Creek: Archaeology, and the 1864 Massacre Site*. Norman: University of Oklahoma Press.

Greene, Victor R. 1964. "A Study of Slavs, Strikes, and Unions: The Anthracite Strike of 1897." *Pennsylvania History* 21: 199–215.

———. 1968. *The Slavic Community on Strike: Immigrant Labor in Pennsylvania Anthracite*. Notre Dame, Ind.: University of Notre Dame Press.

Guydish, Mark. 2015. "Study Ranks NEPA Region as 3rd Worst in Nation for Small Business Employees." *Times Leader*. http://timesleader.com/business/178191/study-ranks-nepa-region-as-3rd-worst-in-nation-for-small-business-employees.

Halbwachs, Maurice. 1992. *On Collective Memory*. Edited and translated by Lewis A. Coser [orig. 1925], Chicago: University of Chicago Press.

Halpin, James. 2014. "Region's Residents Among Most Miserable." *Citizen's Voice*. April 7, 2014.

Hambidge, Jay. 1898. "An Artist's Impression of the Colliery Region." *The Century* 0055(6) April: 822–28. http://digital.library.cornell.edu. Cornell University Library, Making of America.

Handlin, Oscar. 1950. *Race and Nationality in American Life*. Garden City, N.Y.: Doubleday Anchor.

Handlin, Oscar, and Mary Handlin. 1950. "Origins of the Southern Labor System." *William and Mary Quarterly* 7: 199–222.

Harrison, Faye V. 1998. Introduction: Expanding the Discourse on "Race." *American Anthropologist* 100 (3): 609–31.

Hazleton Integration Project. 2016. Hazleton Integration Project. Hazleton One. http://www.hazletonintegrationproject.com/.

Hillyard, Vaughn. 2017. "A More Latino Hazleton, Pa. Braces for Trump's Immigration Orders," *NBC News,* February 1. http://www.nbcnews.com/news/latino/more-latino-hazleton-pa-braces-trump-s-immigration-orders-n714351.

Hine, Chris. 2015. "Joe Maddon Crusades to Save His Hometown." *Chicago Tribune,* February 6. http://www.chicagotribune.com/sports/baseball/cubs/ct-joe-maddon-hazleton-cubs-spt-0208–20150207-story.html.

Hobsbawm, Eric. 1983. "Introduction: Inventing Tradition." In *The Invention of Tradition,* edited by Eric Hobsbawm and Terrence Ranger, pp. 1–14. New York: Cambridge University Press.

Holt, Sharon Ann. 2001. "The Life and Labour of Coxe Miners." *Pennsylvania Legacies* 1 (1) (November): 6–13.

Howe, S. 1999. "Speaking of '98: History, Politics and Memory in the Bicentenary of the 1798 United Irish Uprising." *History Workshop Journal* 47: 222–39.

Hoyt, Henry M. 1898. "Memorandum to Department of State on Facts in Commonwealth vs. James Martin, Sheriff, et al. Court of Oyer and Terminer, Luzerne County, Penn'a, February–March, 1898." In *Papers Relating To The Foreign Relations Of The United States, With The Annual Message of the President Transmitted To Congress December 5, 1898,* pp 97–109. Washington D.C.: U.S. Government Printing Office.

Jones, A. 2007. *Memory and Material Culture.* Cambridge, England: Cambridge University Press.

Jones, Eliott. 1914. *The Anthracite Coal Combination in the United States: With Some Account of the Early Development of the Anthracite Industry.* Cambridge, Mass.: Harvard University Press.

Jordan, Winthrop. 1978. "Unthinking Decision: Enslavement of Negroes in America to 1700." In *Interpreting Colonial America,* edited by Kirby Martin, pp. 140–61. New York: Harper and Row Publishers.

Juravich, Tom. 1985. *Chaos on the Shop Floor.* Philadelphia: Temple University Press.

Kammen, Michael. 1991. *Mystic Chords of Memory: The Transformation of Tradition in American Culture.* New York: Knopf.

———. 1997. "Public History and the Uses of Memory." *Public Historian* 19 (2): 49–52.

Kanjorski, Paul E. 1997. "100th Anniversary of the Lattimer Mine Disaster." House of Representatives, September 10, 1997. http//wwww.thomas.loc.gov/cgi-bin/query/z?r105:E10SE7-111.

Keil, Thomas, and Jacqueline M. Keil. 2015. *Anthracite's Demise and the Post-Coal Economy of Northeastern Pennsylvania.* Bethlehem, Pa.: Lehigh University Press.

Kendree and Shepherd Consultants. 1967a. *Hazle Township Comprehensive Plan: Part One, Development Patterns.* Submitted to Hazle Township Planning Commission. http://elibrary.pacounties.org/Documents/Luzerne_County/1592;%20Hazle%20Township/4207933376mcp.pdf.

———. 1967b. *Hazle Township Comprehensive Plan: Part Two, Development Proposals.* Submitted to Hazle Township Planning Commission. http://elibrary

.pacounties.org/Documents/Luzerne_County/1592;%20Hazle%20Township/4207933376mcp.pdf, accessed December 10, 2014.

Kiernan, John S. 2015. "2015's Best & Worst Cities to Work for a Small Business." *WalletHub.com*. http://wallethub.com/edu/best-cities-to-work-for-a-small-business/3768/#methodology.

Klibanoff, Eleanor. 2015. "The Immigrants it Once Shut Out Bring New Life to Pennsylvania Town." *National Public Radio*, All Things Considered. October 14.

Kohl, Philip L. 1998. "Nationalism and Archaeology: On the Constructions of Nations and the Reconstructions of the Remote Past." *Annual Review of Anthropology* 27: 223–46.

Korson, George. 1938. *Minstrels of the Mine Patch*. Philadelphia: University of Pennsylvania Press.

Kroft, Steve. 2006. "Welcome to Hazleton: One Mayor's Controversial Plan to Deal with Illegal Immigration." 60 Minutes—*CBS News*. http://www.cbsnews.com/stories/2006/11/17/60minutes/main2195789.shtml?tag=contentMain;contentBody.

Kulikoff, Allan. 1986. *Tobacco and Slaves: The Development of Southern Culture in the Chesapeake, 1680–1800*. Chapel Hill: University of North Carolina Press.

"Lattimer Massacre." n.d. Teach PA History. Explore PAhistory.com. http://explorepahistory.com/viewLesson.php?id=1-D-27.

Laws of the General Assembly of the Commonwealth of Pennsylvania Passed at the Session of 1895. 1895. Harrisburg, Pa.: Edwin K. Meyers, State Printer.

Laws of the General Assembly of the Commonwealth of Pennsylvania Passed at the Session of 1895. 1897. Harrisburg, Pa.: Edwin K. Meyers, State Printer.

Linenthal, Edward T. 1993. *Sacred Ground: Americans and Their Battlefields*. Urbana: University of Illinois Press.

Little, Barbara J., and Paul A. Shackel. 2014. *Archaeology, Heritage, and Civic Engagement: Working Toward the Public Good*. Walnut Creek, Calif.: Left Coast Press.

Logan, William. 2012. "Cultural Diversity, Cultural Heritage and Human Rights: Towards Heritage Management as Human Rights-based Cultural Practice." *International Journal of Heritage Studies* 18 (3): 231–44, DOI: 10.1080/13527258.2011.637573.

Longazel, Jamie. 2016. *Undocumented Fears: Immigration and the Politics of Divide and Conquer in Hazleton, Pennsylvania*. Philadelphia: Temple University Press.

Low, Setha. 2009. "Maintaining Whiteness: The Fear of Others and Niceness." *Transforming Anthropology* 17 (2): 79–92.

Lowenthal, David. 1985. *The Past is a Foreign Country*. Cambridge, England: Cambridge University Press.

Ludlow Collective. 2001. "Archaeology of the Colorado Coal Field War, 1913–1914." In *Archaeologies of the Contemporary Past*, edited by Victor Buchli and Gavin Lucas, pp. 94–107. London: Routledge Press.

Luther, P. D. 1881. "Development of Coal Production and Trade in Schuylkill County." In *History of Schuylkill County, PA.*, edited by W. W. Munsell, pp. 41–72. New York: Press of George Macnamara.

MacGaffey, Janet. 2013. *Coal Dust on Your Feet: The Rise, Decline, and Restoration of an Anthracite Mining Town*. Lewisburg, Pa.: Bucknell University Press.

Martin, J. R., and Ruth Wodak. 2003. "Introduction." In *Re/reading the Past: Critical and Functional Perspectives on Time and Value,* edited by J. R. Martin and Ruth Wodak, pp. 1–16. Amsterdam: John Benjamins Publishing.

Matza, Michael. 2016. "10 Years After Immigration Disputes, Hazleton is a Different Place." *Philly.com,* April 4. http://articles.philly.com/2016-04-04/news/72017531_1_amilcar-arroyo-luzerne-county-city-old-coal-city.

McDonald, J. D., L. J. Zimmerman, A. L. McDonald, W. Tall Bull, and T. Rising Sun. 1991. "The Northern Cheyenne Outbreak of 1879: Using Oral History and Archaeology as Tools of Resistance." In *The Archaeology of Inequality,* edited by Randall McGuire and Robert Paynter, pp. 64–78. Oxford, England: Basil Blackwell.

McGlynn, Charles. 1994. Letter, "Charles McGlynn to Stuart Kauffman, Advisor to the Meany Center for Archival Affairs." July 11. On file. Washington, D.C.: Labor Heritage Foundation.

———. 1994. Letter, "Charles McGlynn to Robin Bachin," Newberry Library. July 20. On file. Washington, D.C.: Labor Heritage Foundation.

McGuire, Randall H., and Paul Reckner. 2002. "The Unromantic West: Labor, Capital and Struggle." *Historical Archaeology* 36 (3): 44–58.

McKanders, Karla Mari. 2007. "Welcome to Hazleton! 'Illegal' Immigrants Beware: Local Immigration Ordinances and What the Federal Government Must Do About It." *Loyola University Chicago Law Journal* 39 (1) (2007). University of Tennessee Legal Studies Research Paper No. 65. http://papers.ssrn.com/sol3/papers.cfm?abstract_id=1407727.

Meany, George. 1972. Unidentified news clipping, September 30, 1972. On file, George Meany Memorial Archives 98-002 34/19. College Park: University of Maryland.

Meskell, Lynn M. 2003. "Memory's Materiality: Ancestral Presence, Commemorative Practice and Disjunctive Locales." In *Archaeologies of Memory,* edited by Ruth M. Van Dyke and Susan E. Alcock, 34–55. Malden, Mass.: Blackwell Press.

Miller, Donald L., and Richard E. Sharpless. 1985. *The Kingdom of Coal: Work, Enterprise, and Ethnic Communities in the Mine Fields*. Philadelphia: University of Pennsylvania Press.

Morgan, Edmund. 1975. *American Slavery: American Freedom*. New York: W.W. Norton.

Morgan, Lewis Henry. 1877. *Ancient Society: Researches in the Line of Human Progress from Savagery through Barbarism to Civilization*. New York: Henry Holt.

Muro, Mark, Jonathan Rothwell, Scott Andes, Kenan Fikri, and Siddharth Kulkarni. 2015. *America's Advanced Industries: What They Are, Where They Are, and Why They Matter*. Brookings Institute. https://www.brookings.edu/research/americas-advanced-industries-what-they-are-where-they-are-and-why-they-matter/.

Nora, Pierre. 1999. *Realms of Memory: Rethinking the French Past*. Translated by Arthur Goldhammer. New York: Columbia University Press.

Novak, Michael. 1978. *The Guns of Lattimer*. New York: Basic Books.

O'Bannon, Patrick, Marin Abbot, Susan Nabors, and James Parkinson. 1997. "Anthracite Industry in Northeastern, Pennsylvania, 1769–1945." National Register of Historic Places Multiple Property Documentation Form. Philadelphia: Kise, Frank & Straw.

Omi, Michael, and Howard Winant. 1983. "By the River of Babylon: Race in the United States." *Socialist Review* 13: 31–65.

O'Reilly, Andrew. 2013. "Hazleton: Immigration Debate's Battleground Town Suffers an Identity Crisis." *Fox News Latino*. February 19. http://latino.foxnews.com/latino/news/2013/02/19/hazleton-immigration-debate-battleground-town-suffers-identity-crisis/.

Orser, Charles E., Jr. 2003. *Race and Practice in Archaeological Interpretation*. Philadelphia: University of Pennsylvania Press.

———. 2007. *The Archaeology of Race and Racialization in Historic America*. Gainesville: University Press of Florida.

Palladino, Grace. 2006. *Another Civil War: Labor, Capital, and the State in the Anthracite Regions of Pennsylvania, 1840–1868*. New York: Fordham University Press.

Palmer, Henry W. 1913. *Fifty Years at the Bar and in Politics*. Williamsport, Pa.: Snyder and Bischof.

Patterson, Thomas C., and Frank Spencer. 1994. "Racial Hierarchies and Buffer Races." *Transforming Anthropology* 5 (1&2): 20–27.

Pavalko, Ronald M. 1980. "Racism and the New Immigration: A Reinterpretation of the Assimilation of White Ethnics in American Society." *Sociology* 65 (1): 56–77.

Paynter, Robert. 1989. "The Archaeology of Equality and Inequality." *Annual Review of Anthropology* 18: 369–99.

Pennsylvania AFL-CIO News. 1972. "At Lattimer, Pa.: AFL-CIO Unions Plan Memorial Service." *Pennsylvania AFL-CIO News*. February. George Meany Memorial Archives 98–002 34/19. College Park: University of Maryland.

———. 1972. "75 Years Later: Lattimer." *Pennsylvania AFL-CIO News*, September. George Meany Memorial Archives 98–002 34/19. College Park: University of Maryland.

Pinkowski, Edward. 1950. *Lattimer Massacre*. Philadelphia: Sunshine Press.

———. 1997. *Lattimer Massacre*. http://www.infomatics.sunysb.edu/apa/archives/1997/3958.html.

Pittsburgh Business Times. 2012. "Guide to Pennsylvania Schools Statewide Ranking." *Pittsburgh Business Times*, April 6. http://www2.bizjournals.com/pittsburgh/events/2012/school_guide/index.html.

Powell, Benjamin H. 1980. "The Pennsylvania Anthracite Industry, 1769–1976." *Pennsylvania History* (47): 3–27.

Probert, John. 1998. "Capsule Summary of Efforts to Achieve A Fitting and Proper Memorial for Michael Cheslock." Manuscript in possession of author. June 24.

Putnam, Robert D. 2000. *Bowling Alone: The Collapse and Revival of American Community*. New York: Simon and Schuster.

Rayback, Joseph G. 1966. *A History of American Labor*. New York: Free Press.

Reynolds, Robert L. 1960. "The Coal Kings Come to Judgment." *American Heritage* 11 (3): 61.

Rhone, Rosamond D. 1902. "Anthracite Coal Mines and Mining." *Review of Reviews* (26): 54–63.

Richards, John Stuart. 2002. *Early Coal Mining in the Anthracite Region*. Charleston, S.C.: Acadia Publishing.

Roberts, Peter. 1901. *The Anthracite Coal Industry: A Study of Economic Conditions and Relations of the Cooperative Forces in the Development of the Anthracite Coal Industry of Pennsylvania*. New York: Macmillan.

———. 1970 [1904]. *Anthracite Coal Communities*. New York: Arno Press and the New York Times.

———. 1912. *The New Immigration*. New York: Macmillan.

Rogers, Henry Darwin. 1858. *The Geology of Pennsylvania*, 2 vols. Philadelphia: J.B. Lippincott.

Roller, Michael. 2015. *Migration, Modernity and Memory: The Archaeology of the Twentieth Century in a Northeast Pennsylvania Coal Company Town, 1897–2014*. PhD diss., Department of Anthropology, University of Maryland.

Rood, Henry Edward. 1895. *The Company Doctor: An American Story*. New York: Merriam.

———. 1898. "A Pennsylvania Colliery Village: A Polyglot Community." *The Century*, April 0055 (6): 809–21. http://digital.library.cornell.edu.

Rose, Dan. 1981. *Energy Transition and the Local Community: A Theory of Society Applied to Hazleton, Pennsylvania*. Philadelphia: University of Pennsylvania Press.

Roucek, Joseph S. 1969. "The Image of the Slav in U.S. History and in Immigration Policy." *American Journal of Economics and Sociology* 28 (1): 29–48.

Roy, Andrew. 1907. *A History of the Coal Miners of the United States: From the Development of the Mines to the Close of the Anthracite Strike of 1902*. Columbus, Ohio: J.P. Tauger.

Russell-Ciardi, Maggie. 2008. "The Museum as a Democracy-Building Institution: Reflections on the Shared Journeys Program at the Lower East Side Tenement Museum." *The Public Historian* 30 (1): 39–52.

Sacks, Karen B. 1989. "Toward a Unified Theory of Class, Race, and Gender." *American Ethnologist* 16 (3): 534–50.

Saitta, Dean J. 2007. *The Archaeology of Collective Action*. Gainesville: University Press of Florida.

Scham, Sandra A. 1998. "Mediating Nationalism and Archaeology: A Matter of Trust?" *American Anthropologist* 100 (2): 301–8.

Schooley, Harry B, III. 1977. "The Lattimer Massacre and Trial." *Slovakia* 27: 44–61.

Scott, Douglas D. 2003. "Oral Tradition and Archaeology: Conflict and Concordance Examples from Two Indian War Sites." *Historical Archaeology* 37 (3): 55–65.

———. 2011. "Analysis of Ammunition Components from the Lattimer Massacre Site." Lincoln, Neb.: Conner Consulting.

Scott, Douglas D., and Richard A. Fox Jr. 1987. *Archaeological Insight into the Custer Battle*. Norman: University of Oklahoma Press.

Scott, Douglas D., Richard A. Fox Jr., Melissa A. Connor, and Dick Harmon. 1989. *Archaeological Perspectives on the Battle of the Little Bighorn*. Norman: University of Oklahoma Press.

Scott, James C. 1985. *Weapons of the Weak: Everyday Forms of Peasant Resistance*. New Haven, Conn.: Yale University Press.

Shackel, Paul A., 1996. *Culture Change and the New Technology: An Archaeology of the Early American Industrial Era*. New York: Plenum Press.

———. 2001a. *Myth, Memory and the Making of the American Landscape*. Gainesville: University Press of Florida.

———. 2001b. "Public Memory and the Search for Power in American Historical Archaeology." *American Anthropologist* 102 (3): 655–70.

———. 2003. *Memory in Black and White: Race, Commemoration, and the Post-Bellum Landscape*. Walnut Creek, Calif.: AltaMira Press.

———. 2013. "Reversing the Narrative: Using Archaeology to Challenge the Present and Change the Future." *Historical Archaeology* 47 (3): 1–11.

———. 2016. "The Meaning of Place in the Anthracite Region of Northeastern Pennsylvania." *International Journal of Heritage Studies* 22 (3): 200–13.

Shackel, Paul A., and Michael Roller. 2012. "The Gilded Age Wasn't So Gilded in the Anthracite Region of Pennsylvania." *International Journal of Historical Archaeology* 16 (4): 761–75.

Shackel, Paul A., Michael Roller, and Kristin Sullivan. 2011. "Historical Amnesia and the Lattimer Massacre." *SfAA Newsletter*. May. http://sfaanews.sfaa.net/2011/05/.

Shanks, Michael, and Randall H. McGuire. 1996. "The Craft of Archaeology." *American Antiquity* 61 (1996): 75–88.

Sivilich, Dan. 2011. "Electronic Phase I Archaeological Survey Conducted at: Lattimer 1897 Massacre Site Lattimer Road Lattimer, Luzerne County, Pennsylvania." Monmouth, N.J.: Report prepared for the Lattimer Massacre Project, College Park, Maryland, by Battlefield Restoration and Archaeological Volunteer Organization (BRAVO).

Smedley, Audrey. 1998. "'Race' and the Construction of Human Identity." *American Anthropologist* 100 (3): 690–702.

Solomon, Barbara. 1956. *Ancestors and Immigrants*. Cambridge, Mass.: Harvard University Press.

Steckel, Richard H., and Garrett Senney. 2015. "Historical Origins of a Major Killer: Cardiovascular Disease in the American South." Working Paper 21809. *NBER Working Paper Series*. Cambridge, Mass.: National Bureau of Economic Research. http://www.nber.org/papers/w21809.

Stolarik, M. Mark. 1980. *Slovak Migration from Europe to North America, 1870–1918*. Cleveland: Slovak Institute.

———. 2002. "A Slovak Perspective on the Lattimer Massacre." *Pennsylvania History: A Journal of Mid-Atlantic Studies, Special Issue: The Lattimer Massacre (1897)* 69 (1): 31–41.

Sudarkasa, Niara. 1968. "Racial Classifications: Popular and Scientific." *Science and the Concept of Race* 1: 149–64.

Tarone, L. A. 2004. *We Were Here Once: Successes, Mistakes, & Calamities in Hazleton Area History*. Hazleton, Pa.: Citizen Publishing.

Taylor, George Rogers. 1951. *The Transportation Revolution: 1815–1860*. Vol 4 of the Economic History of the United States. New York: Holt, Rinehart & Winston.

Teski, Marea C., and Jacob J. Climo. 1995. "Introduction." In *The Labyrinth of Memory: Ethnographic Journeys,* edited by Marea C. Teski and Jacob J. Climo, pp. 1–10. Westport, Conn.: Bergin & Garvey.

Trachtenberg, Alexander. 1942. *The History of Legislation for the Protection of Coal Miners in Pennsylvania, 1824–1915*. New York: International Publishers.

Trigger, Bruce. 1989. *A History of Archaeological Thought*. Cambridge, England: Cambridge University Press.

Trouillot, Michel-Rolph. 1995. *Silencing the Past: Power and Production of History*. Boston: Beacon Press.

Turner, George A. 1977. "The Lattimer Massacre and Its Sources." *Slovakia* 27 (50): 9–43.

———. 1984. "Ethnic Responses to the Lattimer Massacre." In *Hard Coal, Hard Times: Ethnicity and Labor in the Anthracite Region,* edited by David L. Salay, pp. 126–52. Scranton, Pa.: Scranton Anthracite Museum Association in cooperation with the Commonwealth of Pennsylvania, Pennsylvania Historical and Museum Commission. Anthracite Museum Press.

———. 1990. "The Lattimer Massacre." The History of Northeast Pennsylvania: The Last 100 Years. Second Annual Proceedings of the Pennsylvania Community College Education Conference Center. October 12.

———. 2002. "The Lattimer Massacre: A Perspective from the Ethnic Community." *Pennsylvania History: A Journal of Mid-Atlantic Studies, Special Issue: The Lattimer Massacre (1897)* 69 (1): 11–30.

United Labor Council of Lower Luzerne & Carbon Counties. 1972. "Lattimer Labor Memorial Year. 75th Anniversary Banquet, Sunday, September 10." United Labor Council of Lower Luzerne & Carbon Counties, AFL-CIO. 1972. Memo, on file, Alice Hoffman Collection. State College: Penn State University Library.

United Mine Workers Journal. 1898. "Since the indictment . . ." *United Mine Workers Journal*. March 17.

———. 1907. "Lattimer Victimt [*sic*] Monument." *United Mine Workers Journal*. September 5.

———. 2007. "110 Years after Lattimer, Hard Coal Miners Still Fighting for Justice: The UMWA in the Anthracite." *United Mine Workers Journal* 118 (5): 3–5.

U.S. Senate. 1911. *Reports of the Immigration Commission*. Vol. 1. 61st Congress, 3rd Session. Abstracts of the Reports of Immigration: With Conclusions and Recommendations and Views of the Minority. Washington, D.C.: Government Printing Office.

U.S. Senate. 1911. *Reports of the Immigration Commission*. Vol. 5. 61st Congress, 3rd Session. Diction of Races or Peoples. Washington, D.C.: Government Printing Office.

U.S. Senate. 1911. *Reports of the Immigration Commission*. Vol. 16. 61st Congress, 3rd Session. Immigrants in Industries: Anthracite Coal Mining. Washington D.C.: Government Printing Office.

Van Dyke, Ruth M., and Susan E. Alcock, editors. 2003. *Archaeologies of Memory*. Malden, Mass.: Blackwell.

Vicino, Thomas. 2013. *Suburban Crossroads: The Fight for Local Control of Immigration Policy*. Lanham, Md.: Lexington Books.

Walker, Mark. 2000. "Labor History at the Ground Level: Colorado Coalfield War Archaeology Project." *Labor's Heritage* 11 (1): 58–75.

Wallace, Anthony F. C. 1987. *St. Clair: A Nineteenth-Century Coal Town's Experience with a Disaster Prone Industry*. New York: Knopf.

Warne, Frank Julian. 1904. *The Slav Invasion and the Mine Workers*. Philadelphia: J.B. Lippincott.

Wiebe, Robert H. 1961. "The Anthracite Coal Strike of 1902: A Record of Confusion." *Mississippi Valley Historical Review* 48 (2): 229–51.

Wilson, Gregory S. 2003. "Deindustrialization, Poverty, and Federal Area Redevelopment in the United States, 1945–1965." In *Beyond the Ruins: The Meanings of Deindustrialization*, edited by Jefferson Cowie and Joseph Heathcott, pp. 181–98. Ithaca, N.Y.: Cornell University Press.

Wood, Margaret. 2002. *Fighting for Our Homes: An Archaeology of Women's Domestic Labor and Social Change in a Working Class, Coal Mining Community, 1900–1930*. PhD diss., Department of Anthropology, Syracuse University.

Wyoming Historical and Geological Society. 1997. "The Lattimer Massacre: A 100th Anniversary Commemoration." Brochure, Wilkes-Barre, Pa.: Wyoming Historical and Geological Society.

Wysocki, Bernard, Jr. 1978. "Review of The Guns of Lattimer, by Michael Novak." *Wall Street Journal,* December 13: 24.

Yellen, Samuel. 1936. *American Labor Struggles*. New York: S.A. Russell.

Zinn, Howard. 2003. *A People's History of the United States: 1492–Present*. New York: HarperCollins.

Newspapers

Citizen's Voice *(Wilkes-Barre, Pa.)*

1984. "Lattimer Disaster Remembered." *Citizen's Voice,* September 10.

Pavloski, Tony. 1993. "Honoring the Martyrs of Lattimer." *Citizen's Voice,* September 9. On file. Wilkes-Barre, Pa.: Luzerne County Historical Society.

Daily Standard *(Hazleton, Pa.)*

1897. "The Strikers: Everything Quiet at the Leheigh and Wilkes-Barre Operations Yesterday." *Daily Standard,* August 19.

1897. "The Strikers: They Unanimously Decide to Reject the Company's Proposition." *Daily Standard,* August 20.

1897. "Trouble at Coleraine. Two Strikes Yesterday—The Alien Tax Law the Cause." *Daily Standard,* August 26.

1897. "Another Strike: The Foreigners in the Employ of A.S. Van Winkle Demand Recognition." *Daily Standard,* August 28.

1897. "The Strike: The Situation Becoming More Serious Hourly." *Daily Standard,* August 31.

1897. "The Strike: The Situation becomes More Serious Hourly." *Daily Standard,* September 1.

1897a. "The Strike: Sheriffs of Three Counties Issue a Proclamation." *Daily Standard,* September 7.

1897b. "Over the Hill: The News of McAdoo, Audenreid and all Other South Side Towns." *Daily Standard,* September 7.

1897. "Yesterday's Butchery. A Mob of Heartless Deputies Fire into a Throng of Marchers and Accomplish Deadly Work. Fifteen Dead and Forty Injured." *Daily Standard,* September 11.

1897. "Citizens' Meeting," *Daily Standard,* September 13.

1897. "Friday's Massacre," *Daily Standard,* September 13.

1897. "The Strike: But Very Little New in the Situation Yesterday." *Daily Standard,* September 17.

1897. "Nearing the End: A Settlement of the Strike is Now in Sight." *Daily Standard,* September 20.

1897. "Strike is Over." *Daily Standard,* September 21.

1897. "Had Another Hearing. Much Additional Evidence Adduced at Both Sessions That Took Place Yesterday." *Daily Standard,* September 23.

1897. "Peace Prevails: Everything is Now Quiet in this Remarkable Region." *Daily Standard,* September 27.

1897. "The Austrian Ambassador." *Daily Standard,* October 1.

1897. "The Austrian Minister Formally Complains to Secretary Sherman." *Daily Standard,* October 15.

1898. "The Trial." *Daily Standard,* February 8.

1898. "The Trial: The Story of the Tragic Affair Repeated Daily." *Daily Standard,* February 11.

1898. "Tragic Tale Retold: Nearly All Give the Same Testimony at Yesterday's Session." *Daily Standard,* February 12.

1898. "Still On: The Trial of Sheriff Martin and His Deputies." *Daily Standard,* February 14.

1898. "Lattimer Horror: Dying Man Shot Thrice By a Deputy Who Stood Over Him." *Daily Standard,* February 18.

1898. " 'A Charging Mob!' That is What Attorney Ferris Calls the Defenseless Men Who were Shot in the Back at Lattimer." *Daily Standard,* February 22.

1898. "Their Day in Court: South Siders Who Were 'Terrorized' Testify in the Martin Trial." *Daily Standard,* February 25.

1898. "Closing Pleas. The Trial of Sheriff Martin and His Deputies Nearly at an End." *Daily Standard,* March 8.

1898. "The Trial is Over: Attorney Palmer and the District Attorney Make Eloquent Pleas." *Daily Standard,* March 9.
1898. "Judge Woodward's Charge." *Daily Standard,* March 10.
1898. "Will Erect a Monument." *Daily Standard,* April 22.
1898. "Memorial Service." *Daily Standard,* September 6.
1898. "Saturday's Demonstration: The Program to be Carried Out by the Toilers." *Daily Standard,* September 9.

Grand Forks Daily Herald *(Grand Forks, N.D.)*

1897. "Another Riot. . . . The Coroner's Jury in the Lattimer Butchery Finally Render a Mixed Verdict—The Jury Does Not Altogether Agree—The Killing Described as Wanton and Unjustifiable." *Grand Forks Daily Herald,* September 28.
1901. "Five of the Greatest Strikes in America." *Grand Forks Daily Herald,* September 8.

Hazleton Sentinel *(Hazleton, Pa.)*

1897. "Strike is Spreading. General Shut Down at Lehigh & Wilksberre Collieries, About 3,000 Men Idle." *Hazleton Sentinel,* August 16.
1897. "Striking Miner's Field Day: The Hungarians and Italians Want to be Paid and Recognized as Are the English Speaking Workers." *Hazleton Sentinel,* August 27.
1897a. "The Strike." *Hazleton Sentinel,* August 31.
1897b. "To Refund the Taxes: One Effect of the Alien Tax Law Decision, The Money will be Paid Back." *Hazleton Sentinel,* August 31.
1897a. "Excitement Prevails. The Strikers on the South Side Take Matters in Hand: They Stop all Mines in the District." *Hazleton Sentinel,* September 2.
1897b. "Uniform Wage Scale." *Hazleton Sentinel,* September 2.
1897. "A Quiet Day. Wild Stories Regarding The Strike Created Excitement. A Squabble in Honey Brook." *Hazleton Sentinel,* September 4.
1897. "The Method Criticized." *Hazleton Sentinel,* September 8.
1897. "Strikers Fired Upon. A Scene of Carnage That Baffles Description. Fifteen More Killed." *Hazleton Sentinel,* September 10.
1897. "Newspaper Options." *Hazleton Sentinel,* September 15.
1897. "Mines Resume Work. Women Take to Marching at Lattimer Quietly Dispersed." *Hazleton Sentinel,* September 20.
1897. "Breaker Burned." *Hazleton Sentinel,* September 22.

The Intelligencer *(Philadelphia)*

Jackson, Peter. 2006. "Hazleton Mayor Says Law Aimed at Illegals, Not Every Immigrant," *The Intelligencer,* October 23: B5.

New York Evening Journal

1898. "What is Crime in Pennsylvania Anyhow?" *New York Evening Journal,* March 10.

1898. "Deputies Ready to Shoot. Men Who Fired at Lattimer on Guard against Revenge. Miners in an Ugly Mood. Coal Companies Put Armed Men on Patrol without Warrant of Law." *New York Evening Journal,* March 16.

1898. "Witness's Houses Burned. Sheporwich Testified for Sheriff Martin at Lattimer Trial." *New York Evening Journal,* March 17.

New York Times

1898. "Lattimer Murder Trials: Attempt to Show that Witnesses Expect to Get Damages in the Event of a Conviction, Kicking a Wounded Man. Judge Woodward Refuses to Admit Testimony Because the Accused Deputy is Not Identified—Men Who were Shot as they Ran." *New York Times,* February 11.

1898. "Lattimer Murder Trials: Morris Engleman Testified that the Deputies Told Him to go into His House for Safety. The Shooting Planned Ahead. A Witness Heard a Deputy Say: 'Let Them Go until They Get to Lattimer and Then We'll Shoot Them' Some Wanton Acts." *New York Times,* February 19.

Appelbaum, Binyamin. 2016. "In City Built by Immigrants, Immigration Is the Defining Political Issue." *New York Times,* October 12. www.nytimes.com/2016/10/13/business/economy/hazleton-pennsylvania-donald-trump-immigrants.html?_r=0.

Philadelphia Inquirer

1897. "Mob Rule up in Luzerne District." *Philadelphia Inquirer,* September 3.

1898. "Lattimer's Dead were Remembered. Solemn Services Held at Hazleton in Memory of the Terrible Loss of Life." *Philadelphia Inquirer,* September 11.

1903. "To Observe Lattimer Day." *Philadelphia Inquirer,* August 8.

1903. "Miners' Day at Lattimer." *Philadelphia Inquirer,* September 8.

Plain Speaker *(Hazleton, Pa.)*

1898. "The Jury Complete." *Plain Speaker,* February 3.

1898. "First Witness." *Plain Speaker,* February 4.

1898. "Fourth Day of Trial." *Plain Speaker,* February 5.

1898. "Trial Drags On. Witness Daniel Ferry Swears Raught Shot Fleeing Striker." *Plain Speaker,* February 11.

1898. "The Martin Trial." *Plain Speaker,* February 12.

1898. "Jury Sees Wounds." *Plain Speaker,* February 18.

1898. "Strikers Had Clubs." *Plain Speaker,* February 23.

1898. "A Day of Oratory: Scarlett and Lenahan Talk to the Jury about the Shooting." *Plain Speaker,* March 8.

1898. "The Verdict Ready," *Plain Speaker*, March 9.

1898. "Labor's Day Out: Several Thousand United Mine Workers Pass in Review Parade. Addresses Follow Demonstration." *Plain Speaker,* September 12.

1898. " 'Big Mary' Dead: Lattimer Woman who was a Prominent Figure in the Strike, a Victim of Heart Disease." *Plain Speaker,* November 11.

1902. "Plans for Lattimer Monument." *Plain Speaker,* March 10.
1902. "For Lattimer Monument Fund." *Plain Speaker,* May 23.

Pottsville Republic *(Pottsville, Pa.)*

Franklin, Erica. 1990. "Obsessed: Lattimer Man Struggles to Keep Memory of Massacred Miners Alive." *Pottsville Republican,* November 1.

The Press *(Philadelphia)*

1898. "Sharp Fight at Martin Trial." *The Press,* February 5.
1898. "Wounded Strikers Show Red Scars." *The Press,* February 10.
1898. "New Version of Riot. Attorney George S. Ferris Describes the Fatal Day to the Jury. Claims was Reign of Terror." *The Press,* February 21.
1898. "Scarlet and Lenahan: Attorneys Begin Their Closing Pleas to the Jury." *The Press,* March 5.
1898. "Expected Verdict." *The Press,* March 9.

Santa Fe Daily New Mexican

1897. "The Coal Strike Not Over. Situation in the Pennsylvania Coal Mining Districts Continues to Create Grave Apprehensions." *Santa Fe Daily New Mexican,* September 20.

Standard Speaker *(Hazleton, Pa.)*

Conrad, Ed. 1997a. "From March to Massacre: Dipple Manor Woman recalls Nightmare of Lattimer Massacre." *Standard Speaker,* September 13: 1–2. On file, Hazleton, Pa.: *Standard Speaker.*
———. 1997b. "Historical Site: Harwood, Where March Began, Gets Marker." *Standard Speaker,* September 13: 1–2. On file, Hazleton, Pa.: *Standard Speaker.*
———. 1997c. "March in the Shoes of 'Martyrs of Lattimer': 300 Walk the Path to Mark 100th Year since the Massacre." *Standard Speaker,* September 13: 1–2. On file, Hazleton, Pa.: *Standard Speaker*
Funk, Richard W. 1997. "Massacre Footnote: Man says Proper Recognition Due." *Standard Speaker,* September 15. News clipping, Alice Hoffman Collection. College Station: Penn State University Library.
Gloman, Chuck. 1989. "DeMarco Raps 'Ethnic Slurs' on Monument." *Standard Speaker,* September 26: 9.
Klemlow, Bob. 2009. "Don't Play to Fear; Accept all Hazletonians." *Standard Speaker,* October 7: A10.
McGlynn, Charles. 1992. "Great Debt Owed to Martyrs of Lattimer." *Standard Speaker,* September 2. News clipping on file. Wilkes-Barre, Pa.: Luzerne County Historical Society.
———. 1994. "Lattimer Massacre 97th Anniversary Mass Sept. 9." *Standard Speaker,* August 29. News clipping, on file. Washington, D.C.: Labor Heritage Foundation.

Pinkowski, Edward. 1993. "Lattimer Massacre Martyrs Have Names." *Standard Speaker,* May 21. News clipping, on file. Washington, D.C.: Labor Heritage Foundation.

Ragan, Tom. 2013. "Service Planned to Remember Lattimer Massacre." *Standard Speaker,* August 21.

Tarone, L. A. 1997. "Remembering Lattimer: Bishop Reminds Those at Mass: Remember What Happened Here." *Standard Speaker,* September 14. On file, news clipping Hazleton, Pa.: *Standard Speaker.*

Whalen, Jill. 2010. "Annual Tradition Marking Lattimer Massacre Ends." *Standard Speaker,* September 10.

Wittig, Daniel. 1993. "Hazleton Suffers from Racism." Editorial, *Standard Speaker,* April 28. On file, Washington, D.C.: Labor Heritage Foundation.

Sun *(Baltimore)*

1897. "Strikers Shot: Twenty Killed, Fifteen Fatally Wounded and Nearly Forty Less Severely Hurt." *Sun,* September 11.

Sunday Independent *(Wilkes-Barre, Pa.)*

1937. "County's Bloodiest Union Strike: 19 were Dead after Lattimer Riot in 1897." *Sunday Independent,* June 20.

Wilkes-Barre Times *(Wilkes-Barre, Pa.)*

1903. "Labor Notes: The Lattimer Monument." *Wilkes-Barre Times,* September 16.

1907. "Monument to Lattimer Men: Will be Erected in West Hazleton to Cost $5,000." *Wilkes-Barre Times,* August 24.

1907. "The Lattimer Monument: Mine Workers' Officials Meeting at Hazleton To-Day to Take Final Action." *Wilkes-Barre Times,* October 15.

1908. "Design Selected for Lattimer Monument: Fitting Testimony to Memory of Men Who Died for Their Cause." *Wilkes-Barre Times,* April 17.

1908. "Lattimer's Eleventh Anniversary." *Wilkes-Barre Times,* September 11.

Times Leader *(Wilkes-Barre, Pa.)*

Kester, Marcella. 2016. "Cubs Coach Joe Maddon Continues Thanksmas Tradition In Hometown of Hazleton." *Times Leader,* December 18. https://timesleader.com/sports/616945/.

Learn-Andes, Jennifer. 2004. "Amazon's Queen." *Times Leader,* January 11. News clipping on file. Wilkes-Barre, Pa.: Osterhout Free Library.

Moisey, Linda. 1993. "Mass Observes Lattimer Massacre." *Times Leader,* September 11. On file, Wilkes-Barre, Pa.: Luzerne County Historical Society.

Wall Street Journal *(New York)*

Jordan, Miriam. 2006. "Grassroots Groups Boost Clout in Immigration Fight." *Wall Street Journal,* September 28. http://www.post-gazette.com/pg/06271/725845-84.stm.

Wilkes-Barre Semi-Weekly Record *(Wilkes-Barre, Pa.)*

1903. "Monument Site: the One to the Memory of the Lattimer Victims" *Wilkes-Barre Semi-Weekly Record,* July 31: 7. On file. Wilkes-Barre, Pa.: Luzerne County Historical Society.

Wilkes-Barre Times Leader *(Wilkes-Barre, Pa.)*

1911. "Miners' Officials Decide on Extensive Rebuilding Plan: Expect President White Sunday. Monument for Victims of Lattimer Massacre and Other Important Questions Discussed by District Boards at Hazleton To-Day." *Wilkes-Barre Times Leader,* September 22.

1912. "Monument for the Lattimer Victims is Again Proposed." *Wilkes-Barre Times Leader,* September 18.

Letters

Brennan, John C., to Michael Novak. 1976, April 9. Letter on file, Series No. 4.1.1, Guns of Lattimer. Michael Novak Papers. Easton, Mass.: Stonehill College Archives and Historical Collections.

Pinkowski, Edward, to Michael Novak. 1973, January 12. Letter on file, Series No. 4.1.1, Guns of Lattimer. Michael Novak Papers. Easton, Mass.: Stonehill College Archives and Historical Collections.

Interviews

Moye, John. 1973. March 13. Series no. 4.1.1, Guns of Lattimer. Michael Novak Papers. Easton, Mass.: Stonehill College Archives and Historical Collections.

Stabert, Anna. 1973. January 17. Series no. 4.1.1, Guns of Lattimer. Michael Novak Papers. Easton, Mass.: Stonehill College Archives and Historical Collections.

Tomasko, John. 1977. August 6. Series no. 4.1.1, Guns of Lattimer. Michael Novak Papers. Easton, Mass.: Stonehill College Archives and Historical Collections.

INDEX

Paul A. Shackel is a professor in the Department of Anthropology at the University of Maryland–College Park. His books include *Archaeology, Heritage and Civic Engagement: Working Toward the Public Good* and *New Philadelphia: An Archaeology of Race in the Heartland*.

THE WORKING CLASS IN AMERICAN HISTORY

Worker City, Company Town: Iron and Cotton-Worker Protest in Troy and Cohoes, New York, 1855–84 *Daniel J. Walkowitz*

Life, Work, and Rebellion in the Coal Fields: The Southern West Virginia Miners, 1880–1922 *David Alan Corbin*

Women and American Socialism, 1870–1920 *Mari Jo Buhle*

Lives of Their Own: Blacks, Italians, and Poles in Pittsburgh, 1900–1960 *John Bodnar, Roger Simon, and Michael P. Weber*

Working-Class America: Essays on Labor, Community, and American Society *Edited by Michael H. Frisch and Daniel J. Walkowitz*

Eugene V. Debs: Citizen and Socialist *Nick Salvatore*

American Labor and Immigration History, 1877–1920s: Recent European Research *Edited by Dirk Hoerder*

Workingmen's Democracy: The Knights of Labor and American Politics *Leon Fink*

The Electrical Workers: A History of Labor at General Electric and Westinghouse, 1923–60 *Ronald W. Schatz*

The Mechanics of Baltimore: Workers and Politics in the Age of Revolution, 1763–1812 *Charles G. Steffen*

The Practice of Solidarity: American Hat Finishers in the Nineteenth Century *David Bensman*

The Labor History Reader *Edited by Daniel J. Leab*

Solidarity and Fragmentation: Working People and Class Consciousness in Detroit, 1875–1900 *Richard Oestreicher*

Counter Cultures: Saleswomen, Managers, and Customers in American Department Stores, 1890–1940 *Susan Porter Benson*

The New England Working Class and the New Labor History *Edited by Herbert G. Gutman and Donald H. Bell*

Labor Leaders in America *Edited by Melvyn Dubofsky and Warren Van Tine*

Barons of Labor: The San Francisco Building Trades and Union Power in the Progressive Era *Michael Kazin*

Gender at Work: The Dynamics of Job Segregation by Sex during World War II *Ruth Milkman*

Once a Cigar Maker: Men, Women, and Work Culture in American Cigar Factories, 1900–1919 *Patricia A. Cooper*

A Generation of Boomers: The Pattern of Railroad Labor Conflict in Nineteenth-Century America *Shelton Stromquist*

Work and Community in the Jungle: Chicago's Packinghouse Workers, 1894–1922 *James R. Barrett*

Workers, Managers, and Welfare Capitalism: The Shoeworkers and Tanners of Endicott Johnson, 1890–1950 *Gerald Zahavi*

Men, Women, and Work: Class, Gender, and Protest in the New England Shoe Industry, 1780–1910 *Mary Blewett*

Workers on the Waterfront: Seamen, Longshoremen, and Unionism in the 1930s *Bruce Nelson*

German Workers in Chicago: A Documentary History of Working-Class Culture from 1850 to World War I *Edited by Hartmut Keil and John B. Jentz*

On the Line: Essays in the History of Auto Work *Edited by Nelson Lichtenstein and Stephen Meyer III*

Labor's Flaming Youth: Telephone Operators and Worker Militancy, 1878–1923 *Stephen H. Norwood*

Another Civil War: Labor, Capital, and the State in the Anthracite Regions of Pennsylvania, 1840–68 *Grace Palladino*

Coal, Class, and Color: Blacks in Southern West Virginia, 1915–32 *Joe William Trotter Jr.*

For Democracy, Workers, and God: Labor Song-Poems and Labor Protest, 1865–95 *Clark D. Halker*

Dishing It Out: Waitresses and Their Unions in the Twentieth Century *Dorothy Sue Cobble*

The Spirit of 1848: German Immigrants, Labor Conflict, and the Coming of the Civil War *Bruce Levine*

Working Women of Collar City: Gender, Class, and Community in Troy, New York, 1864–86 *Carole Turbin*

Southern Labor and Black Civil Rights: Organizing Memphis Workers *Michael K. Honey*

Radicals of the Worst Sort: Laboring Women in Lawrence, Massachusetts, 1860–1912 *Ardis Cameron*

Producers, Proletarians, and Politicians: Workers and Party Politics in Evansville and New Albany, Indiana, 1850–87 *Lawrence M. Lipin*

The New Left and Labor in the 1960s *Peter B. Levy*

The Making of Western Labor Radicalism: Denver's Organized Workers, 1878–1905 *David Brundage*

In Search of the Working Class: Essays in American Labor History and Political Culture *Leon Fink*

Lawyers against Labor: From Individual Rights to Corporate Liberalism *Daniel R. Ernst*

"We Are All Leaders": The Alternative Unionism of the Early 1930s *Edited by Staughton Lynd*

The Female Economy: The Millinery and Dressmaking Trades, 1860–1930 *Wendy Gamber*

"Negro and White, Unite and Fight!": A Social History of Industrial Unionism in Meatpacking, 1930–90 *Roger Horowitz*

Power at Odds: The 1922 National Railroad Shopmen's Strike
Colin J. Davis

The Common Ground of Womanhood: Class, Gender, and Working Girls' Clubs, 1884–1928 *Priscilla Murolo*

Marching Together: Women of the Brotherhood of Sleeping Car Porters *Melinda Chateauvert*

Down on the Killing Floor: Black and White Workers in Chicago's Packinghouses, 1904–54 *Rick Halpern*

Labor and Urban Politics: Class Conflict and the Origins of Modern Liberalism in Chicago, 1864–97 *Richard Schneirov*

All That Glitters: Class, Conflict, and Community in Cripple Creek *Elizabeth Jameson*

Waterfront Workers: New Perspectives on Race and Class
Edited by Calvin Winslow

Labor Histories: Class, Politics, and the Working-Class Experience
Edited by Eric Arnesen, Julie Greene, and Bruce Laurie

The Pullman Strike and the Crisis of the 1890s: Essays on Labor and Politics *Edited by Richard Schneirov, Shelton Stromquist, and Nick Salvatore*

AlabamaNorth: African-American Migrants, Community, and Working-Class Activism in Cleveland, 1914–45 *Kimberley L. Phillips*

Imagining Internationalism in American and British Labor, 1939–49 *Victor Silverman*

William Z. Foster and the Tragedy of American Radicalism
James R. Barrett

Colliers across the Sea: A Comparative Study of Class Formation in Scotland and the American Midwest, 1830–1924 *John H. M. Laslett*

"Rights, Not Roses": Unions and the Rise of Working-Class Feminism, 1945–80 *Dennis A. Deslippe*

Testing the New Deal: The General Textile Strike of 1934 in the American South *Janet Irons*

Hard Work: The Making of Labor History *Melvyn Dubofsky*

Southern Workers and the Search for Community: Spartanburg County, South Carolina *G. C. Waldrep III*

We Shall Be All: A History of the Industrial Workers of the World (abridged edition) *Melvyn Dubofsky, ed. Joseph A. McCartin*

Race, Class, and Power in the Alabama Coalfields, 1908–21 *Brian Kelly*

Duquesne and the Rise of Steel Unionism *James D. Rose*

Anaconda: Labor, Community, and Culture in Montana's Smelter City *Laurie Mercier*

Bridgeport's Socialist New Deal, 1915–36 *Cecelia Bucki*

Indispensable Outcasts: Hobo Workers and Community in the American Midwest, 1880–1930 *Frank Tobias Higbie*

After the Strike: A Century of Labor Struggle at Pullman *Susan Eleanor Hirsch*
Corruption and Reform in the Teamsters Union *David Witwer*
Waterfront Revolts: New York and London Dockworkers, 1946–61 *Colin J. Davis*
Black Workers' Struggle for Equality in Birmingham *Horace Huntley and David Montgomery*
The Tribe of Black Ulysses: African American Men in the Industrial South *William P. Jones*
City of Clerks: Office and Sales Workers in Philadelphia, 1870–1920 *Jerome P. Bjelopera*
Reinventing "The People": The Progressive Movement, the Class Problem, and the Origins of Modern Liberalism *Shelton Stromquist*
Radical Unionism in the Midwest, 1900–1950 *Rosemary Feurer*
Gendering Labor History *Alice Kessler-Harris*
James P. Cannon and the Origins of the American Revolutionary Left, 1890–1928 *Bryan D. Palmer*
Glass Towns: Industry, Labor, and Political Economy in Appalachia, 1890–1930s *Ken Fones-Wolf*
Workers and the Wild: Conservation, Consumerism, and Labor in Oregon, 1910–30 *Lawrence M. Lipin*
Wobblies on the Waterfront: Interracial Unionism in Progressive-Era Philadelphia *Peter Cole*
Red Chicago: American Communism at Its Grassroots, 1928–35 *Randi Storch*
Labor's Cold War: Local Politics in a Global Context *Edited by Shelton Stromquist*
Bessie Abramowitz Hillman and the Making of the Amalgamated Clothing Workers of America *Karen Pastorello*
The Great Strikes of 1877 *Edited by David O. Stowell*
Union-Free America: Workers and Antiunion Culture *Lawrence Richards*
Race against Liberalism: Black Workers and the UAW in Detroit *David M. Lewis-Colman*
Teachers and Reform: Chicago Public Education, 1929–70 *John F. Lyons*
Upheaval in the Quiet Zone: 1199/SEIU and the Politics of Healthcare Unionism *Leon Fink and Brian Greenberg*
Shadow of the Racketeer: Scandal in Organized Labor *David Witwer*
Sweet Tyranny: Migrant Labor, Industrial Agriculture, and Imperial Politics *Kathleen Mapes*
Staley: The Fight for a New American Labor Movement *Steven K. Ashby and C. J. Hawking*
On the Ground: Labor Struggles in the American Airline Industry *Liesl Miller Orenic*

NAFTA and Labor in North America *Norman Caulfield*
Making Capitalism Safe: Work Safety and Health Regulation in America, 1880–1940 *Donald W. Rogers*
Good, Reliable, White Men: Railroad Brotherhoods, 1877–1917 *Paul Michel Taillon*
Spirit of Rebellion: Labor and Religion in the New Cotton South *Jarod Roll*
The Labor Question in America: Economic Democracy in the Gilded Age *Rosanne Currarino*
Banded Together: Economic Democratization in the Brass Valley *Jeremy Brecher*
The Gospel of the Working Class: Labor's Southern Prophets in New Deal America *Erik Gellman and Jarod Roll*
Guest Workers and Resistance to U.S. Corporate Despotism *Immanuel Ness*
Gleanings of Freedom: Free and Slave Labor along the Mason-Dixon Line, 1790–1860 *Max Grivno*
Chicago in the Age of Capital: Class, Politics, and Democracy during the Civil War and Reconstruction *John B. Jentz and Richard Schneirov*
Child Care in Black and White: Working Parents and the History of Orphanages *Jessie B. Ramey*
The Haymarket Conspiracy: Transatlantic Anarchist Networks *Timothy Messer-Kruse*
Detroit's Cold War: The Origins of Postwar Conservatism *Colleen Doody*
A Renegade Union: Interracial Organizing and Labor Radicalism *Lisa Phillips*
Palomino: Clinton Jencks and Mexican-American Unionism in the American Southwest *James J. Lorence*
Latin American Migrations to the U.S. Heartland: Changing Cultural Landscapes in Middle America *Edited by Linda Allegro and Andrew Grant Wood*
Man of Fire: Selected Writings *Ernesto Galarza, ed. Armando Ibarra and Rodolfo D. Torres*
A Contest of Ideas: Capital, Politics, and Labor *Nelson Lichtenstein*
Making the World Safe for Workers: Labor, the Left, and Wilsonian Internationalism *Elizabeth McKillen*
The Rise of the Chicago Police Department: Class and Conflict, 1850–1894 *Sam Mitrani*
Workers in Hard Times: A Long View of Economic Crises *Edited by Leon Fink, Joseph A. McCartin, and Joan Sangster*
Redeeming Time: Protestantism and Chicago's Eight-Hour Movement, 1866–1912 *William A. Mirola*

Struggle for the Soul of the Postwar South: White Evangelical Protestants and Operation Dixie *Elizabeth Fones-Wolf and Ken Fones-Wolf*
Free Labor: The Civil War and the Making of an American Working Class *Mark A. Lause*
Death and Dying in the Working Class, 1865–1920 *Michael K. Rosenow*
Immigrants against the State: Yiddish and Italian Anarchism in America *Kenyon Zimmer*
Fighting for Total Person Unionism: Harold Gibbons, Ernest Calloway, and Working-Class Citizenship *Robert Bussel*
Smokestacks in the Hills: Rural-Industrial Workers in West Virginia *Louis Martin*
Disaster Citizenship: Survivors, Solidarity, and Power in the Progressive Era *Jacob A. C. Remes*
The Pew and the Picket Line: Christianity and the American Working Class *Edited by Christopher D. Cantwell, Heath W. Carter, and Janine Giordano Drake*
Conservative Counterrevolution: Challenging Liberalism in 1950s Milwaukee *Tula A. Connell*
Manhood on the Line: Working-Class Masculinities in the American Heartland *Steve Meyer*
On Gender, Labor, and Inequality *Ruth Milkman*
The Making of Working-Class Religion *Matthew Pehl*
Civic Labors: Scholar Activism and Working-Class Studies *Edited by Dennis Deslippe, Eric Fure-Slocum, and John W. McKerley*
Victor Arnautoff and the Politics of Art *Robert W. Cherny*
Against Labor: How U.S. Employers Organized to Defeat Union Activism *Edited by Rosemary Feurer and Chad Pearson*
Teacher Strike! Public Education and the Making of a New American Political Order *Jon Shelton*
Hillbilly Hellraisers: Federal Power and Populist Defiance in the Ozarks *J. Blake Perkins*
Sewing the Fabric of Statehood: Garment Unions, American Labor, and the Establishment of the State of Isreal *Adam Howard*
Labor and Justice across the America *Edited by Leon Fink and Juan Manuel Palacio*
Frontiers of Labor: Comparative Histories of the United States and Australia *Edited by Greg Patmore and Shelton Stromquist*
Women Have Always Worked: A Concise History, Second Edition *Alice Kessler-Harris*
Remembering Lattimer: Labor, Migration, and Race in Pennsylvania Anthracite Country *Paul A. Shackel*

The University of Illinois Press
is a founding member of the
Association of American University Presses.

University of Illinois Press
1325 South Oak Street
Champaign, IL 61820-6903
www.press.uillinois.edu